WORKSHEETS
FOR CLASSROOM OR LAB PRACTICE

CARRIE GREEN

ELEMENTARY AND INTERMEDIATE ALGEBRA
THIRD EDITION

Tom Carson

Bill Jordan
Seminole State College of Florida

Addison-Wesley
is an imprint of

Reproduced by Pearson Addison-Wesley from electronic files supplied by the author.

Copyright © 2011, 2007, 2003 Pearson Education, Inc.
Publishing as Pearson Addison-Wesley, 75 Arlington Street, Boston, MA 02116.

ISBN-13: 978-0-321-62730-8
ISBN-10: 0-321-62730-X

1 2 3 4 5 6 BB 14 13 12 11 10

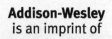

Addison-Wesley
is an imprint of

PEARSON

www.pearsonhighered.com

Table of Contents

The **Worksheets for Classroom or Lab Practice** is a workbook that was designed to be used in a variety of ways. We see it as an organizational tool; a self- and teacher-assessment tool; a place to get more practice, to show work, and to keep notes. The first two pages are designated for taking notes, defining key terms and procedures, explaining topics in students' own words, and working through guided examples that correlate to the examples in the text. Each guided example presents all the steps of a solution. Students must complete each example by either entering the appropriate math in the space provided or by filling in the missing annotations to explain in their own words why a particular step is necessary. The remaining pages provide many practice problems to assess students' understanding of the topics at hand. The odd answers to these practice problems are provided at the back of the supplement, and a complete set of answers is available to instructors on-line.

Chapter 1   FOUNDATIONS OF ALGEBRA

1.1 Number Sets and the Structure of Algebra

KEY VOCABULARY

Term	Definition	Example
Variable		
Constant		
Expression		
Equation		
Inequality		
Set		
Rational number		
Irrational number		
Real number		
Absolute value		

GUIDED EXAMPLES

1. Graph $6\frac{1}{4}$ on a number line.

    Solution

    The number $6\frac{1}{4}$ is located _____ of the way between the whole numbers

    _____ and _____ . Divide the space between _____ and _____ into

    _____ equal divisions and place a dot on the _____ mark.

2. Simplify $\left|-3.7\right|$.

    Solution

    $\left|-3.7\right| =$ _____ because -3.7 is _____ units from 0 on a number line.

3. Use =, <, or > to write a true statement.

    10 _____ -12

    Solution

    The completed statement is 10 _____ -12 because 10 is farther to the [left / right] on

    a number line than -12.

NOTES

2

PRACTICE PROBLEMS

Write a set representing each description.

1. The set of all letters in the word "dance"

1._____

2. The set of all even natural numbers less than 5

2._____

Determine whether each number is a rational number or an irrational number.

3. $\dfrac{5}{7}$

3._____

4. $\sqrt{14}$

4._____

5. -0.4

5._____

6. $0.\overline{345}$

6._____

Graph each number on a number line.

7. $\dfrac{5}{2}$

7.

$\longleftarrow\!\longrightarrow$

8. $-3\dfrac{1}{2}$

8.

$\longleftarrow\!\longrightarrow$

9. 4.6

9.

$\longleftarrow\!\longrightarrow$

Simplify.

10. $\left|-25\right|$

10._____

3

11. $\left|\dfrac{1}{4}\right|$

Use =, <, or > to write a true statement.

12. -14 _____ -8

13. 3.9 _____ $|-3.9|$

List the given numbers in order from least to greatest.

14. $7.8,\ -5.5,\ \left|-4\dfrac{1}{4}\right|,\ 0,\ -14.1,\ |-1.9|$

15. $0,\ -12.9,\ |-1.6|,\ 7.4,\ -6.1,\ \left|-2\dfrac{1}{4}\right|$

4

Chapter 1   FOUNDATIONS OF ALGEBRA

1.2 Fractions

KEY VOCABULARY

Term	Definition	Example
Fraction		
Multiple		
LCM		
LCD		
Factors		
Prime number		
Prime factorization		
Lowest terms		

KEY PROPERTIES, PROCEDURES, OR STRATEGIES

Eliminating a Common Factor in a Fraction

In the Language of Math	In Your Own Words

Simplifying a Fraction to Lowest Terms

In the Language of Math	In Your Own Words

GUIDED EXAMPLE
Simplify to lowest terms.

$$\frac{24}{28}$$

Solution

$\frac{24}{28} =$ 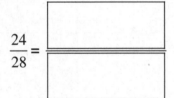   **Replace the numerator and denominator with their prime factorizations.**

$=$ 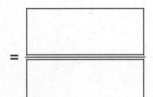   **Eliminate the common prime factors.**

$=$ 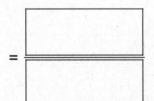   **Simplify.**

6

PRACTICE PROBLEMS

Identify the fraction represented by each shaded region.

1.

1._____

2.

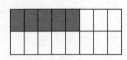

2._____

Find the missing number that makes the fractions equivalent.

3. $\dfrac{3}{7} = \dfrac{?}{28}$

3._____

4. $-\dfrac{5}{11} = \dfrac{?}{99}$

4._____

5. $-\dfrac{42}{66} = \dfrac{7}{?}$

5._____

Write the fractions as equivalent fractions with the LCD.

6. $\dfrac{7}{15}$ and $\dfrac{11}{12}$

6._____

7. $-\dfrac{11}{21}$ and $-\dfrac{7}{18}$

7._____

7

8. $\dfrac{3}{14}$ and $\dfrac{7}{12}$

8._____

Write the prime factorization for each number.

9. 96

9._____

10. 100

10._____

11. 525

11._____

Simplify to lowest terms.

12. $-\dfrac{30}{36}$

12._____

13. $\dfrac{25}{35}$

13._____

14. $\dfrac{56}{72}$

14._____

15. $-\dfrac{12}{20}$

15._____

Chapter 1   FOUNDATIONS OF ALGEBRA

1.3 Adding and Subtracting Real Numbers; Properties of Real Numbers

KEY VOCABULARY

Term	Definition	Example
Additive inverses		

KEY PROPERTIES, PROCEDURES, OR STRATEGIES

Additive Identity

In the Language of Math	In Your Own Words

Commutative Property of Addition

In the Language of Math	In Your Own Words

Associative Property of Addition

In the Language of Math	In Your Own Words

Adding Numbers with the Same Sign

In the Language of Math	In Your Own Words

Name:

Date:

Instructor:

Section:

Adding Numbers with Different Signs

In the Language of Math	In Your Own Words

Adding Fractions

In the Language of Math	In Your Own Words

Rewriting Subtraction

In the Language of Math	In Your Own Words

GUIDED EXAMPLE

Subtract.

$$\frac{1}{10} - \left(-\frac{2}{3}\right)$$

Solution

$$\frac{1}{10} - \left(-\frac{2}{3}\right)$$

 **Write the subtraction as an equivalent addition.**

 **Write equivalent fractions with a common denominator.**

 **Simplify.**

10

PRACTICE PROBLEMS

Indicate whether each equation illustrates the additive identity, commutative property of addition, associative property of addition, or additive inverse.

1. $0.9 + (-0.9) = 0$ **1.**_____

2. $0.2 + 0 = 0.2$ **2.**_____

3. $-11 + (5 + 12) = (-11 + 5) + 12$ **3.**_____

Add.

4. $-26 + 62$ **4.**_____

5. $\dfrac{2}{3} + \dfrac{5}{6}$ **5.**_____

6. $-46.8 + (-4.2)$ **6.**_____

Find the additive inverse.

7. 8 **7.**_____

8. $-\dfrac{1}{2}$ **8.**_____

9. a **9.**_____

Simplify.

10. $-(-15)$ **10.**_____

11. $-|7|$

11._____

12. $-|-92|$

12._____

Subtract.

13. $-5-6$

13._____

14. $-3-(-4)$

14._____

15. $1.6-5.4$

15._____

Chapter 1   FOUNDATIONS OF ALGEBRA

1.4 Multiplying and Dividing Real Numbers; Properties of Real Numbers

KEY VOCABULARY

Term	Definition	Example
Multiplicative inverses		

KEY PROPERTIES, PROCEDURES, OR STRATEGIES

Properties of Multiplication	In the Language of Math	In Your Own Words
Multiplicative Property of 0		
Multiplicative Identity		
Commutative Property of Multiplication		
Associative Property of Multiplication		
Distributive Property of Multiplication over Addition		
Multiplying Signed Numbers		
Multiplying Fractions		
Multiplying Decimal Numbers		

Properties of Division	In the Language of Math	In Your Own Words
Dividing Signed Numbers		
Division Involving 0		
Dividing Fractions		
Dividing Decimal Numbers		

GUIDED EXAMPLE

Multiply.

$$\frac{7}{3} \cdot \left(-\frac{1}{28}\right)$$

Solution

$$\frac{7}{3} \cdot \left(-\frac{1}{28}\right) = \boxed{}$$ **Divide out the common factor.**

$$= \boxed{}$$ **There is an [even / odd] number of negative factors, so the product is [positive / negative].**

14

PRACTICE PROBLEMS

Indicate whether each equation illustrates the multiplicative property of 0, the multiplicative identity, the commutative property of multiplication, the associative property of multiplication, or the distributive property.

1. $7 \cdot (8 \cdot 56) = (7 \cdot 8) \cdot 56$

1._____

2. $9(x + y) = 9x + 9y$

2._____

3. $1 \cdot \left(-\dfrac{3}{4} \right) = -\dfrac{3}{4}$

3._____

Multiply.

4. $-7 \cdot (-4) \cdot (-7)$

4._____

5. $\dfrac{2}{7} \cdot \left(-\dfrac{1}{2} \right)$

5._____

6. $-\dfrac{11}{9} \cdot \dfrac{3}{10}$

6._____

7. $(-6.2)(6.8)$

7._____

Find the multiplicative inverse.

8. $\dfrac{4}{5}$

8._____

9. 8   9._____

Divide.

10. $-18 \div (-6)$ 10._____

11. $7 \div 0$ 11._____

12. $-\dfrac{6}{5} \div \dfrac{3}{2}$ 12._____

13. $7.2 \div 0.12$ 13._____

Solve.

14. Planet A has an average surface temperature of   14._____
$-150°$ F. Planet B has an average surface
temperature that is $\dfrac{3}{2}$ times that of planet A. Find
the average surface temperature on planet B.

15. In a poll where respondents can agree, disagree, or   15._____
have no opinion, $\dfrac{5}{14}$ of the respondents said they
agreed and $\dfrac{4}{5}$ of those that agreed were women.
What fraction of all respondents were women who
agreed with the statement in the poll?

16

Chapter 1   FOUNDATIONS OF ALGEBRA

1.5 Exponents, Roots, and Order of Operations

KEY VOCABULARY

Term	Definition	Example
Exponent		
Base		

KEY PROPERTIES, PROCEDURES, OR STRATEGIES

Evaluating an Exponential Form

In the Language of Math	In Your Own Words

Properties of Square Roots	In the Language of Math	In Your Own Words
Square Roots Involving the Radical Sign		
Square Root of a Product or Quotient		
Square Root of a Sum or Difference		

Name: Date:
Instructor: Section:

Order-of-Operations Agreement

Finding the Arithmetic Mean

In the Language of Math	In Your Own Words

GUIDED EXAMPLE

Simplify using the order of operations.

$$4^2 - 135 \div (19 - 4)$$

Solution

$$4^2 - 135 \div (19 - 4)$$

=    **Calculate within the parentheses.**

=    **Evaluate the exponential form.**

=   **Perform the division.**

=   **Subtract.**

PRACTICE PROBLEMS

Evaluate.

1. 2^8

1._____

2. $(-7)^3$

2._____

3. $(0.28)^3$

3._____

Find all square roots of each number.

4. 144

4._____

5. 289

5._____

Evaluate the square roots.

6. $\sqrt{0.64}$

6._____

7. $\sqrt{-25}$

7._____

8. $\sqrt{\dfrac{49}{25}}$

8._____

Simplify using the order of operations.

9. $19 \cdot 26 + 44$

9._____

10. $5 - (-4)(-7)^2$

10._____

11. $5.9 + (2.2)^2 - 9(7-9)$

11._____

12. $-7|6-8| + 2^2$

12._____

13. $-30 \div (-3)(8) + \sqrt{169 - 144} + 14$

13._____

14. $-36 \cdot \dfrac{2}{3} \div (-4) + |9 - 4(5+2)|$

14._____

Solve.

15.   Tomeka has the following test scores in a history course.
83, 85, 96, 81, 96, 86, 82
What is the average of her test scores?

15._____

20

Chapter 1   FOUNDATIONS OF ALGEBRA

1.6 Translating Word Phrases to Expressions

Operation	Variable Expression	Word Phrases
Addition		
Subtraction		
Multiplication		
Division		

GUIDED EXAMPLES

Translate each phrase to an algebraic expression.

    a)   the difference of some number and nine

Solution

_____

Select a variable to represent the unknown number.

Subtraction is not commutative, so translate the subtraction in the correct order.

    b)   the sum of thirty-four and z

Solution

_____

The commutative property of addition allows us to write the expression in either order.

    c)   the ratio of twelve to y, all raised to the fifth power

Solution

_____

When coupled with the word *ratio*, the word *to* translates to the fraction line.

The word *all* indicates that we should use parentheses to group the expression.

    d)   A driver drove at a speed of 51 mph for y hours. Write an algebraic expression for the distance the driver drove.

Solution

_____

Distance equals rate times time.

NOTES

22

● **PRACTICE PROBLEMS**

Translate each phrase to an algebraic expression.

1. eight times c **1.**_____

2. six more than two times a number   **2.**_____

3. seventy-five increased by d **3.**_____

4. the ratio of twelve to a nonzero number y **4.**_____

● **5.** five less than three times a number·   **5.**_____

6. a divided by nine   **6.**_____

7. one-fourth subtracted from the product of two and y **7.**_____

8. the product of negative four and the difference of a number and one   **8.**_____

23

9. a number minus seven times the difference between the number and sixteen

9._____

10. the difference of ten and *b*, all raised to the fourth power

10._____

11. the quotient of six and a number, decreased by eight

11._____

12. the product of two and a number, increased by nine

12._____

13. negative nineteen increased by the sum of *a* and *b*

13._____

14. a number minus four times the sum of the number and sixty

14._____

15. The length of a rectangle is four more than the width. If the width is represented by *x*, write an algebraic expression that describes the length.

15._____

Chapter 1   FOUNDATIONS OF ALGEBRA

1.7 Evaluating and Rewriting Expressions

KEY VOCABULARY

Term	Definition	Example
Terms		
Coefficient		
Like terms		

KEY PROPERTIES, PROCEDURES, OR STRATEGIES

Evaluating an Algebraic Expression

In the Language of Math	In Your Own Words

Combining Like Terms

In the Language of Math	In Your Own Words

GUIDED EXAMPLES

1. Evaluate $\left| 2x^2 - 3y \right|$ when $x = 3$ and $y = 6$.

Solution

$\left| 2x^2 - 3y \right|$

$=$ [   ]   **Replace x with 3 and y with 6.**

$=$ [   ]   **Simplify the exponential form.**

$=$ [   ]   **Multiply.**

$=$ [   ]   **Subtract.**

$=$ [   ]   **Find the absolute value.**

2. Combine like terms.
$1.2p + 0.4q - 0.18p + 0.2q$

Solution

$1.2p + 0.4q - 0.18p + 0.2q$

$=$ [   ] $+$ [   ]   **Collect the like terms.**

$=$ [   ]   **Combine like terms.**

NOTES

PRACTICE PROBLEMS

Evaluate the expressions using the given values.

1. $4(c+6)-5d$; $c=5$, $d=4$

1._____

2. $3x^2-4x+7$; $x=-4$

2._____

3. $-7\sqrt{x}+4\sqrt{y}$; $x=36$, $y=9$

3._____

Determine all values that cause each expression to be undefined.

4. $\dfrac{6}{y-4}$

4._____

5. $\dfrac{9}{6y+5}$

5._____

Use the distributive property to write an equivalent expression and simplify.

6. $4(c+7)$

6._____

7. $-2(b-5)$

7._____

Identify the coefficient of each term.

8. $-8w^5$ **8.**_____

9. $-x^7$ **9.**_____

10. $\dfrac{c}{3}$ **10.**_____

Simplify by combining like terms.

11. $3d - 18d$ **11.**_____

12. $4.7x - 8.9x$ **12.**_____

13. $3p + 13 - 20p + 34$ **13.**_____

14. $-36 - 4x + 3y - 93 - y + 8x$ **14.**_____

15. $\dfrac{20}{11}r + \dfrac{3}{11}s - \dfrac{5}{11}r + \dfrac{16}{11}s$ **15.**_____

Chapter 2   SOLVING LINEAR EQUATIONS AND INEQUALITIES

2.1 Equations, Formulas, and the Problem-Solving Process

KEY VOCABULARY

Term	Definition	Example
Equation	A problem with an equal sign	dad + house = x
Solution	The answer to a problem	$x = 5$
Formula		
Perimeter	The outline of anything	
Area	The measurement of the inside of anything	
Volume	The area of a 3D object	
Circumference	the perimeter of a circle.	
Radius	the imiginary line from the mid. pt. to the end	
Diameter	The line from 2 ends of the circle	

KEY PROPERTIES, PROCEDURES, OR STRATEGIES

Checking a possible solution

Problem-Solving Outline

Calculating the Area of Composite Figures

GUIDED EXAMPLE

Check to see if the given value is a solution to the equation.

$$6.7n - 3.09 = 2.6n + 9.21; \; n = 3$$

Solution

$$6.7n - 3.09 = 2.6n + 9.21$$

$$6.7(3) - 3.09 \overset{?}{=} 2.6(3) + 9.21$$

Replace n with 3 and see if the equation is true.

$$\boxed{} \overset{?}{=} \boxed{}$$

$$\boxed{} \overset{?}{=} \boxed{}$$

3 [is / is not] a solution to the equation.

PRACTICE PROBLEMS

Check to see if the given number is a solution to the given equation.

1. $5x + 7 = 30$; $x = 5$

1. _false_

2. $7(y-3) = -2y + 60$; $y = 10$

2. _false_

3. $6.3u - 4.13 = 2.6u + 3.64$; $u = 2.1$

3. _false_

4. $p^4 - 5p = p^3 + 5$; $p = -6$

4. _____

5. $|v^2 - 81| = -v + 9$; $v = 9$

5. _true_

31

Solve using geometric formulas.

6.   A contractor needs to put up a wallpaper border around a rectangular room. The room is 12 feet by 22 feet.

     a. What is the total length of wallpaper border needed?

     b. If the wallpaper border comes in packages of 12 feet, how many packages are needed to finish the project?

     c. What is the total cost of the border if the packages are priced at $12.66 each?

6a.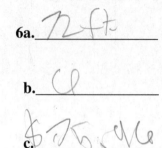

b. _____

c. _____

7.   A circular tunnel is used to accelerate elementary particles. The radius of the tunnel is 7 kilometers. If a particle travels one complete revolution around the circumference of the tunnel, what distance does it travel? Use 3.14 for π. Round to the nearest tenth of a kilometer if needed.

7. _____

8.   One of the large rectangular rooms in a hotel has dimensions 52 feet by 119 feet. What is the total square footage of the floor of the room?

8. _____

9.   A flight departs from city A at 7:00 A.M. EST to arrive in city B at 1:30 P.M. PST. If the plane flies an average rate of $330\frac{1}{4}$ mph, what distance does it travel, rounded to the nearest tenth of a mile? Use the formula relating distance, rate, and time: $d = rt$. (Hint: There is a three-hour time difference between EST and PST.)

9. _____

Chapter 2   SOLVING LINEAR EQUATIONS AND INEQUALITIES

2.2 The Addition Principle of Equality

KEY VOCABULARY

Term	Definition	Example
Linear equation		
Linear equation in one variable		
Identity		
Contradiction		

KEY PROPERTIES, PROCEDURES, OR STRATEGIES

The Addition Principle of Equality

In the Language of Math	In Your Own Words

Using the Addition Principle of Equality

33

Solving Linear Equations

GUIDED EXAMPLE

Solve and check.

$$4.2q - 3.7 - 4.8q = -1.6q - 5.5$$

    Solution

$$4.2q - 3.7 - 4.8q = -1.6q - 5.5$$

Simplify the expressions; then isolate q.

Combine $4.2q$ and $-4.8q$.

Add $1.6q$ to both sides.

Add 3.7 to both sides to isolate q.

The solution is $q =$

Check: $4.2q - 3.7 - 4.8q = -1.6q - 5.5$

Replace q in the original equation with the solution and verify that the equation is true.

Is the equation true?

Name: Date:
Instructor: Section:

PRACTICE PROBLEMS

Determine whether each equation is linear.

1. $7x + 7 = 5x - 4$ 1._____

2. $6x^2 + x^6 = 18$ 2._____

3. $13x - 11y = 12$ 3._____

Solve and check.

4. $-13 = x + 14$ 4._____

5. $x - \dfrac{3}{10} = -\dfrac{3}{5}$ 5._____

6. $8(y + 7) - 7y = 16 - 8$ 6._____

7. $1.3(m + 8) - (6.6 + 0.3m) = -9 + 6$ 7._____

8. $6x = 7x - 1$ 8._____

9. $-8p - 3 = -11 - 7p$ 9._____

10. $10x + 4 + 4x + 8 = 13x + 3$ 10._____

11. $25 + w + 8w - 10 = 15w + 19 - 6w - 4$ 11._____

12. $10.5 - 0.9z - 1.2 - 1.2z = 13 - 2.1z - 7.1$ 12._____

Translate to an equation and then solve.

13. Dorota knows the distance from her home to work   13._____
is 34 miles. Unfortunately, she gets a flat tire 18
miles from work. How far did Dorota drive before
her tire went flat?

14. Robert is playing a dice game. To earn points, he   14._____
needs to correctly add up the total value of five dice
he rolls on each turn. If he has a total of 22 on four
of the dice, what does the fifth die need to be so that
his score will be 26?

Chapter 2   SOLVING LINEAR EQUATIONS AND INEQUALITIES

2.3 The Multiplication Principle of Equality

KEY PROPERTIES, PROCEDURES, OR STRATEGIES

The Multiplication Principle of Equality

In the Language of Math	In Your Own Words

Using the Multiplication Principle of Equality

Solving Linear Equations

GUIDED EXAMPLE

Solve the following linear equation for the indicated variable and check the solution.

$$3(m-10)+4(m-18)=10$$

Solution

$$3(m-10)+4(m-18)=10$$

Distribute to clear parentheses.

Combine like terms.

Use the addition principle so that all variable terms are on one side of the equation and all constants are on the other side.

Combine like terms.

Use the multiplication principle to clear any remaining coefficient.

The solution is $m =$ [ ]

Check: $3(m-10)+4(m-18)=10$

Replace m in the original equation with the solution and verify that the equation is true.

Is the equation true? [ ]

38

Name: Date:
Instructor: Section:

PRACTICE PROBLEMS

Solve and check.

1. $9x = 90$

1._____

2. $-2x = -44$

2._____

3. $\dfrac{x}{6} = 9$

3._____

4. $\dfrac{1}{14}p = \dfrac{7}{6}$

4._____

5. $-5x + 6 = 41$

5._____

6. $\dfrac{1}{2}y + 6 = -11$

6._____

7. $8x - (3x + 7) = 28$

7._____

8. $2(y-2)-1=3(y-3)$ 8._____

9. $\dfrac{2}{3}+\dfrac{1}{2}t=\dfrac{1}{3}$ 9._____

10. $\dfrac{8}{5}(m+1)=\dfrac{3}{5}-m$ 10._____

11. $0.8y-0.3y=7.5$ 11._____

12. $0.2(b-12)=0.6b$ 12._____

Solve for the unknown amount.

13. A truck's cargo area contains 1638 cubic feet of   13._____
space. If the cargo area is 7 feet wide and 9 feet
high, how long must it be? (Use $V=lwh.$)

14. A right circular cylinder has radius 4 inches and   14._____
volume 48π cubic inches. Find the height. (Use
$V=\pi r^2 h.$)

40

Chapter 2   SOLVING LINEAR EQUATIONS AND INEQUALITIES

2.4 Applying the Principles to Formulas

KEY PROPERTIES, PROCEDURES, OR STRATEGIES

Isolating a Variable in a Formula

GUIDED EXAMPLES

1. Isolate d in the formula $\dfrac{4}{5} + 9g = \dfrac{d}{h}$.

    Solution

    $$\dfrac{4}{5} + 9g = \dfrac{d}{h}$$

    To isolate d, we must clear h. Because h is dividing d, we

    _____ both sides by h.

    Simplify.

2. Isolate t in the formula $\dfrac{Q}{t} = \pi$.

    Solution

    $$\dfrac{Q}{t} = \pi$$

    To isolate t, we must get it out of the denominator. Multiply both sides by t. Then clear π.

    Simplify.

3. Isolate *m* in the formula $f = \dfrac{1}{4}mv$.

Solution

$$f = \frac{1}{4}mv$$

To isolate *m*, we must clear $\dfrac{1}{4}$ and *v*. Because $\dfrac{1}{4}$ and *v* are multiplying *m*, we

_____ both sides by $\dfrac{1}{4}$ and *v*. (You can do this in one step or in two steps.)

Simplify.

NOTES

PRACTICE PROBLEMS

Solve for the indicated variable.

1. $v - 9b = m; v$ 1._____

2. $3x - 4 = n; x$ 2._____

3. $10x + 9y = 3; y$ 3._____

4. $\dfrac{rs}{19} - C = t; C$ 4._____

5. $9(s + 5z) = g - jh; s$ 5._____

6. $\dfrac{c}{9} + \dfrac{d}{7} = 5; d$ 6._____

7. $b = \dfrac{S - Q}{ny}; n$ 7._____

43

8. $P = a + 2b + 4c; a$ 8._____

9. $C = \dfrac{5}{3}\pi g^4; g^4$ 9._____

10. $b = \dfrac{1}{2}gr; g$ 10._____

11. $U = \dfrac{x}{9}(c + p); p$ 11._____

12. $A = y + ytk; t$ 12._____

13. $Q = \dfrac{7}{4}M + 21; M$ 13._____

14. $q = 5g + 5v; g$ 14._____

Chapter 2   SOLVING LINEAR EQUATIONS AND INEQUALITIES

2.5 Translating Word Sentences to Equations

Key Words for an Equal Sign

Translating Algebraic Equations to English Sentences

Equation	Translation
	Equations Involving Addition
$x + 8 = 15$	
	Equations Involving Subtraction
$z - 4 = 11$	
	Equations Involving Multiplication
$0.9y = 15$	
	Equations Involving Division
$\dfrac{a}{7} = 10$	
	Equations That Involve More Than One Operation
$6m + 6 = 2$	
	Equations Involving Parentheses
$4(c - 3) = 12$	

GUIDED EXAMPLE

Translate the following sentence to an equation and then solve.

The quotient of two less than a number and ten is the same as the number divided by twenty.

Solution

Understand   The key word *quotient* indicates _____ and the key

words *less than* indicate _____. The key words *is*

the same as indicate _____ and the key words

divided by indicate _____.

Plan   Use the key words to translate to an equation and then solve. Use *x* for the variable.

Execute   Translate:

The quotient of	two less than a number	and ten	is the same as	the number	divided by	twenty.

Solve:

Answer $x =$ [ ]

Check

PRACTICE PROBLEMS

Translate each sentence to an equation and then solve.

1. A number multiplied by five is negative twenty.   1._____

2. Sixteen less than a number is twelve.   2._____

3. A number divided by ten is seven-elevenths.   3._____

4. One-half of a number is five-sixths.   4._____

5. Four more than the product of five and *x* yields   5._____
    fifty-four.

6. Twelve less than four times a number is twelve.   6._____

7. Three times the sum of b and six is equal to negative twenty-four.

_____

8. One-fourth of the sum of a number and two is seven.

_____

9. Fifteen less than nine times a number is equal to that number added to one.

_____

10. Six is the result when eight is subtracted from the ratio of a number to ten.

10._____

11. The difference of a number and fifteen subtracted from the difference of twice the number and nine is three.

11._____

12. The sum of eleven and four times t is the same as the difference of six times t and seventeen.

12._____

Chapter 2   SOLVING LINEAR EQUATIONS AND INEQUALITIES

2.6 Solving Linear Inequalities

KEY VOCABULARY

Term	Definition	Example
Linear inequality		

KEY PROPERTIES, PROCEDURES, OR STRATEGIES

The Addition Principle of Inequality

In the Language of Math	In Your Own Words

The Multiplication Principle of Inequality

In the Language of Math	In Your Own Words

Graphing Inequalities

49

Solving Linear Inequalities

GUIDED EXAMPLE

Solve $4+5x \leq -7x-12$ and write the solution set in set-builder notation and interval notation; then graph the solution set.

Solution

Use the addition principle to separate the variable terms and constant terms; then clear the remaining coefficient.

$4+5x \leq -7x-12$

Add 7x to both sides.

Subtract 4 from both sides.

Divide both sides by the coefficient of x.

Solution

Set-builder notation:

Interval notation:

Graph: ←—————————————→

Name: Date:
Instructor: Section:

PRACTICE PROBLEMS

Write the solution set in set-builder notation and interval notation, then graph the solution set.

1. $x \leq -1$

1._____

_____

<----------------------------------->

2. $y > 4$

2._____

_____

<----------------------------------->

3. $-2 < n \leq 3$

3._____

_____

<----------------------------------->

For each graph, write the inequality in set-builder notation and interval notation.

4.

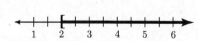

4._____

_____

5.

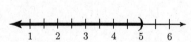

5._____

_____

6.

6._____

_____

51

Name: Date:

Instructor: Section:

Solve each inequality. Write the solution set in set-builder notation and interval notation. Graph the solution set.

7. $7 - 9x < 61$ **7.**_____

_____

⟵⎯⎯⎯⎯⎯⎯⎯⎯⎯⟶

8. $7 - w < 16 + w$ **8.**_____

_____

⟵⎯⎯⎯⎯⎯⎯⎯⎯⎯⟶

9. $4(4k + 2) - 8(k - 2) \geq 3(2k + 3) - 1$ **9.**_____

_____

⟵⎯⎯⎯⎯⎯⎯⎯⎯⎯⟶

Translate to an inequality and then solve.

10. Twelve more than a number is greater than thirty-eight.   **10.**_____

11. Three times a number less twenty-six is at least fifty-five.   **11.**_____

12. The difference of sixteen times a number and ten is less than five times the number.   **12.**_____

Solve.

13. Jacqui has grades of 90 and 80 on her first two algebra tests. If she wants an average of at least 71, what possible scores can she make on her third test?   **13.**_____

52

Chapter 3   PROBLEM SOLVING

3.1 Ratios and Proportions

KEY VOCABULARY

Term	Definition	Example
Ratio		
Unit Ratio		
Proportion		
Congruent angles		
Similar figures		

KEY PROPERTIES, PROCEDURES, OR STRATEGIES

Proportions and Their Cross Products

In the Language of Math	In Your Own Words

Solving a Proportion

Solving Proportion Application Problems

GUIDED EXAMPLE

Solve for the missing number in the proportion.

$$\frac{-4}{3} = \frac{16}{x}$$

Solution

$$\frac{-4}{3} = \frac{16}{x}$$

____ · ____ = ____   ____ · ____ = ____   **Calculate the cross products.**

$\boxed{}$ = $\boxed{}$ **Set the cross products equal to each other.**

$\boxed{}$ = $\boxed{}$ **Divide both sides by the coefficient of x.**

The solution is $\boxed{}$

NOTES

PRACTICE PROBLEMS

Write the ratio in simplest form.

1. One molecule of methanol contains 1 carbon atom, 4 hydrogen atoms, and 1 oxygen atom. Write the ratio of carbon atoms to total atoms in the molecule.

    1._____

2. The back wheel of a bicycle rotates $1\frac{2}{7}$ times with $1\frac{3}{11}$ rotations of the pedals. Write the ratio of back wheel rotations to pedal rotations in simplest form.

    2._____

Write each as a unit ratio.

3. Oscar drove 95 miles and used 5 gallons of gas. What is the unit ratio of miles to gallons? Interpret the answer.

    3._____

4. The same kind of sparkling water comes in two types of cartons.
    Carton A: Six 10-oz bottles for $3.99
    Carton B: Four 12-oz bottles for $2.39
    Which carton is the better buy?

    4._____

Determine whether the ratios are equal.

5. $\frac{14}{15} \overset{?}{=} \frac{42}{45}$

    5._____

55

6. $\dfrac{9.6}{6.3} \overset{?}{=} \dfrac{48.0}{25.2}$

6._____

7. $\dfrac{1\frac{1}{5}}{4\frac{1}{4}} \overset{?}{=} \dfrac{4\frac{4}{5}}{12\frac{3}{4}}$

7._____

Solve for the missing number.

8. $\dfrac{24}{23} = \dfrac{x}{4.6}$

8._____

9. $\dfrac{-14}{n} = \dfrac{12}{48}$

9._____

Solve.

10. A car travels 887 kilometers in 13 days. At this rate, how far would it travel in 39 days?

10._____

Find the missing lengths in the similar figures.

11.

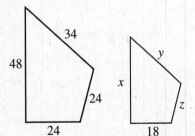

11._____

Chapter 3   PROBLEM SOLVING

3.2 Percents

KEY VOCABULARY

Term	Definition	Example
Percent		

KEY PROPERTIES, PROCEDURES, OR STRATEGIES

Rewriting a Percent

Writing a Fraction or Decimal as a Percent

Translating Simple Percent Sentences

Solving Percent Application Problems

GUIDED EXAMPLES

1. Write 45% as a decimal and as a fraction in simplest form.

    Solution

    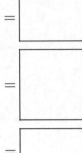

    $45\% = \dfrac{\boxed{}}{100}$ **Write as a ratio with 100 in the denominator.**

    $= \boxed{}$ **Write the decimal form.**

    $= \boxed{}$ **Write the fraction form.**

    $= \boxed{}$ **Simplify to lowest terms.**

2. 60% of what number is 48?

    Solution

    Note the three pieces:

60%	of	what number	is	48?
↑		↑		↑
Percent	of	the whole	is	the part.

    Method 1.   Word-for-word translation:

    60% of what number is 48?

    $\quad\quad\downarrow \quad\quad \downarrow \quad\quad \downarrow$

    $0.60 \quad \cdot \quad n \quad = 48$

    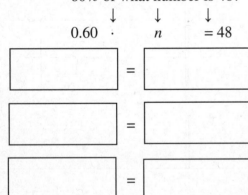

    Method 2. Proportion:

    $\text{Percent} = \dfrac{\text{Part}}{\text{Whole}}$

    $\dfrac{\boxed{}}{100} = \dfrac{\boxed{}}{\boxed{}}$ ← Part
    ← Whole

    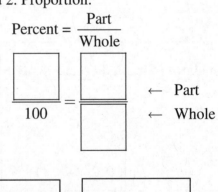

58

PRACTICE PROBLEMS

Write each percent as a decimal and as a fraction in simplest form.

1. 65%

1._____

2. 9.25%

2._____

3. $14\frac{1}{3}\%$

3._____

Write as a percent rounded to the nearest tenth if necessary.

4. $\frac{4}{6}$

4._____

5. 0.05

5._____

6. $\frac{5}{9}$

6._____

Translate word for word or translate to a proportion; then solve.

7. 70% of 410 is what number?

7._____

8. What is $75\frac{1}{2}\%$ of $8610?

8._____

9. 51.85 is 61% of what number?

9._____

10. What percent of 88 is 22?

10._____

Solve.

11. Mary earns $340 per week and has 25% of this amount withheld for taxes, Social Security, and Medicare. Find the amount withheld.

11._____

12. A football player made 659 field goals of 1018 attempts. What percent of his field goal attempts did he make? Round to the nearest tenth of a percent.

12._____

13. The sales tax rate in Kentucky is 6%. How much tax is charged on a purchase of 4 telephones at $39 apiece? What is the total price?

13._____

14. During a sale, a dress decreased in price from $90 to $63. What was the percent decrease in the price of the dress?

14._____

15. Due to a slump in the economy, a mutual fund has dropped by 30% from last year to this year. If the fund is now worth $12,495, how much was the fund worth last year?

15._____

Chapter 3   PROBLEM SOLVING

3.3 Problems with Two or More Unknowns

KEY VOCABULARY

Term	Definition	Example
Complementary angles		
Supplementary angles		

KEY PROPERTIES, PROCEDURES, OR STRATEGIES

Solving Problems with Two or More Unknowns

NOTES

GUIDED EXAMPLE

A textbook has 20 more than three times the number of pages as a workbook. Combined, they have 200 pages. Find the number of pages in each.

Solution

Understand There are two unknowns in the problem. We must find:

_____ and

_____.

Relationship 1: The textbook has 20 more than three times the number of pages in the workbook.

Relationship 2: The total number of pages in the textbook and the workbook is 200.

Plan Translate the relationships to an equation and then solve.

Execute Use the first relationship to determine which unknown will be represented by a variable and represent the other unknown in terms of that variable.

Relationship 1: The textbook has 20 more than three times the number of pages in the workbook.

Number of pages in the workbook: _____

Number of pages in the textbook: _____

Now use the second relationship to write an equation.

Relationship 2: The total number of pages in the textbook and the workbook is 200.

| Number of pages in the workbook | + | Number of pages in the textbook | = 200 |

[ ] + [ ] = 200

[ ] = 200   **Combine like terms.**

[ ] = [ ]   **Isolate the variable.**

[ ] = [ ]

Answer Number of pages in the workbook: [ ]

Number of pages in the textbook: [ ]

Check Verify that the total number of pages in both books is 200.

PRACTICE PROBLEMS

Translate to an equation and then solve.

1. Amanda is six times as old as Michael. In 16 years, Amanda will be only twice as old as Michael. What are their ages now?

1._____

2. A Special Olympics event has 4 more boys than girls competing. The total number of participants is 1000. How many boys competed and how many girls competed?

2._____

3. The length of a rectangular mailing label is 3 centimeters less than twice the width. The perimeter is 30 centimeters. Find the dimensions of the label.

3._____

4. Two angles are supplementary. One is $90°$ more than twice the other. Find the measures of the angles.

4._____

5.   The sum of three consecutive even integers is 180.   5._____
What are the integers?

6.   The sum of three consecutive integers is 186. What   6._____
are the integers?

7.   Find three consecutive even integers such that twice   7._____
the smallest plus four times the largest will result in
the middle integer plus 84.

Complete a table, write an equation, then solve.

8.   The school's fall musical was a big success. For   8._____
opening night, 347 tickets were sold. Students paid
$3.50 each, while non-students paid $5.50 each. If a
total of $1450.50 was collected, how many students
and how many non-students attended?

Chapter 3   PROBLEM SOLVING

3.4 Rates

KEY PROPERTIES, PROCEDURES, OR STRATEGIES

Two Objects Traveling in Opposite Directions

Two Objects Traveling in the Same Direction

NOTES

GUIDED EXAMPLE

A freight train leaves a station and travels north at 60 miles per hour. Four hours later, a passenger train leaves on a parallel track and travels north at 100 miles per hour. How long will it take the passenger train to overtake the freight train? How far from the station will they meet?

Solution

Understand To determine the time it takes for the passenger train to overtake the freight train, use a table to organize the rates and times.

Let t represent the time for the passenger train to catch up to the freight train. Add 4 hours to t to represent the freight train's time. Use the formula $d = rt$.

Categories	Rate	Time	Distance
Freight train			
Passenger train			

Plan   Set the expressions for the individual distances equal and solve for t.

Execute   Freight train's distance = Passenger train's distance

Length of time the freight train has traveled: _____ hours

Distance the freight train has traveled: _____ miles

Length of time the passenger train has traveled: _____ hours

Distance the passenger train has traveled: _____ miles

Answer   It will take the passenger train _____ hours to catch up to the freight

train. At that time, the trains have each traveled _____ miles.

Check   Verify that the trains have traveled the same distance after each train has traveled its respective length of time.

66

PRACTICE PROBLEMS

The following exercises involve people or objects moving in opposite directions. Complete a table, write an equation, and then solve.

1. Adam left Cleveland traveling north at an average speed of 61 miles per hour. Two hours later Kris left Cleveland traveling south at an average speed of 58 miles per hour. How long will it take after Adam left for them to be 955 miles apart?

    1._____

2. A train leaves Dallas and travels north at 85 kilometers per hour. Another train leaves at the same time and travels south at 70 kilometers per hour. How long will it take before they are 620 kilometers apart?

    2._____

3. Two cars pass each other traveling in opposite directions. One car is going $1\frac{1}{2}$ times as fast as the other car. At the end of $5\frac{1}{2}$ hours, they are 440 miles apart. Find each car's rate.

    3._____

The following exercises involve people or objects moving in the same direction. Complete a table, write an equation, and then solve.

4. A truck enters a highway driving 60 miles per hour. A car enters the highway at the same place 8 minutes later and drives 68 miles per hour in the same direction. From the time the car enters the highway, how many minutes will it take the car to pass the truck?

4._____

5. Two cars start driving in the same direction from the same place. If one travels 52 miles per hour and the other 61 miles per hour, how long will it take them to be 63 miles apart?

5._____

6. Allison drove for 2 hours on the freeway, then decreased her speed by 20 miles per hour and drove 4 more hours on a country road. If her total trip was 268 miles, then what was her speed on the freeway?

6._____

68

Chapter 3   PROBLEM SOLVING

3.5 Investment and Mixture

KEY PROPERTIES, PROCEDURES, OR STRATEGIES

Solving Problems Involving Investment

Solving Mixture Problems

NOTES

Name: Date:

Instructor: Section:

GUIDED EXAMPLE

A pharmacist has a 40% acid solution and a 25% acid solution. How many liters of each must be mixed to form 105 liters of a 27% acid solution?

Solution

Understand There are three solutions in this problem: the 40% solution, the 25% solution, and the new 27% solution that is created by mixing the 40% and 25% solutions. Create a table that has a row for each solution.

We are to find the volume of the 40% solution and the volume of the 25% solution. Choose a variable to represent one of these unknown amounts, and use the fact that we want to form 105 liters of a 27% acid solution to represent the other unknown amount.

Solutions	Concentration of Acid	Volume of Solution	Volume of Acid
40% solution			
25% solution			
27% solution			

Plan   Write an equation that describes the mixture. Then solve for the variable.

Execute   Volume of acid   Volume of acid   Volume of acid
               in the   +   in the   =   in the
          40% solution   25% solution   27% solution

Answer   Volume of 40% acid solution needed: _____ liters

                Volume of 25% acid solution needed: _____ liters

Check   Verify that the volume of acid in the two original solutions combined is equal to the volume of acid in the combined solution.

70

PRACTICE PROBLEMS

The following exercises involve investment. Complete a table, write an equation, and then solve.

1. Kate invests money in two plans. She invests three-fifths of the money in an account at a return rate of 4%. She invests the remainder of the money in an account with a return rate of 3%. If the total interest earned in one year from the investments is $57.60, how much was invested in each plan?

1._____

2. A mother wants to invest $15,000 for her son's future education. She invests a portion of the money in a bank certificate of deposit (CD account) which earns 4% and the remainder in a savings bond that earns 7%. If the total interest earned after one year is $900, how much money was invested in the CD account?

2._____

3. Larry has two investments totaling $20,000. Plan A has an APR of 3% and plan B has an APR of 11%. After one year, the interest earned from plan B is $940 more than the interest from plan A. How much did he invest in each plan?

3._____

The following exercises involve mixtures. Complete a table, write an equation, and then solve.

4. How many gallons of a 70% antifreeze solution must be mixed with 70 gallons of 10% antifreeze to get a mixture that is 60% antifreeze?

4._____

5. A chocolate factory makes dark chocolate that is 33% fat and a white chocolate that is 47% fat. How many kilograms of dark chocolate should be mixed with 700 kilograms of white chocolate to make a ripple blend that is 35% fat?

5._____

6. Keil is going to make 13 pounds of mixed nuts for a party. Peanuts cost $3.00 per pound and fancy nuts cost $6.00 per pound. If Keil can spend $63.00 on nuts, how many pounds of each should he buy?

6._____

Chapter 4   GRAPHING LINEAR EQUATIONS AND INEQUALITIES

4.1 The Rectangular Coordinate System

KEY PROPERTIES, PROCEDURES, OR STRATEGIES

Identifying the Coordinates of a Point

Plotting a Point

Identifying Quadrants

Name: Date:
Instructor: Section:

GUIDED EXAMPLE
Write the coordinates for each point shown.

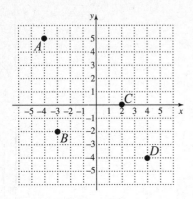

Solution

A: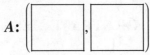

From point A, a vertical line intersects the x-axis at _____ and a horizontal line intersects the y-axis at _____.

B: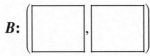

From point B, a vertical line intersects the x-axis at _____ and a horizontal line intersects the y-axis at _____.

C: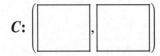

From point C, a vertical line intersects the x-axis at _____ and a horizontal line intersects the y-axis at _____.

D: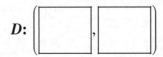

From point D, a vertical line intersects the x-axis at _____ and a horizontal line intersects the y-axis at _____.

NOTES

Name: Date:
Instructor: Section:

PRACTICE PROBLEMS

Write the coordinates for each point.

1.

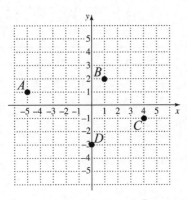

1._____

_____

_____

_____

2.

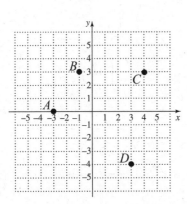

2._____

_____

_____

_____

Plot and label the points indicated by the coordinate pairs.

3. $(-4,-1)$, $(2,5)$, $(-2,-4)$, $(5,0)$

3.

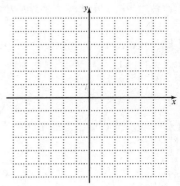

4. $(-2,5)$, $(0,1)$, $(1,-5)$, $(5,5)$

4.

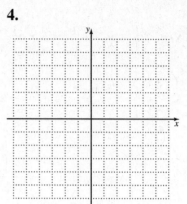

State the quadrant in which the point is located. If the point lies on an axis, state which axis.

5. $(7.2,-112)$

5._____

6. $(0,708)$

6._____

7. $(-65,301)$

7._____

8. $(-34,0)$

8._____

9. $(45,267)$

9._____

10. $(-58,-0.26)$

10._____

Determine whether the set of points is linear or nonlinear.

11. (age of tree, number of growth rings):
    $(1,15)$, $(2,20)$, $(10,30)$, $(20,100)$

11._____

12. (age of child, height):
    $(2,32)$, $(3,36)$, $(4,40)$, $(5,44)$

12._____

13. (age of child, ounces of milk):
    $(1,8)$, $(3,6)$, $(5,6)$, $(7,10)$

13._____

Chapter 4   GRAPHING LINEAR EQUATIONS AND INEQUALITIES

4.2 Graphing Linear Equations

KEY PROPERTIES, PROCEDURES, OR STRATEGIES

Checking a Potential Solution for an Equation with Two Variables

Finding Solutions to Linear Equations with Two Variables

Graphing Linear Equations

Horizontal Lines

In the Language of Math	In Your Own Words

Vertical Lines

In the Language of Math	In Your Own Words

GUIDED EXAMPLES

1. Determine whether the ordered pair is a solution for the equation.

    $(2, -4); 9x - 9y = 15$

    Solution

    $$9x - 9y = 15$$

    $$9(\quad) - 9(\quad) \overset{?}{=} 15 \qquad \textbf{Replace } x \textbf{ with 2 and } y \textbf{ with } -4.$$

    $$(\quad) - (\quad) \overset{?}{=} 15$$

    $$\underline{\qquad} \overset{?}{=} 15 \qquad \textbf{The ordered pair [is / is not] a solution for the equation.}$$

2. Find three solutions for the equation $3x + 2y = 6$.

    Solution

    To find a solution, we replace one of the variables with a chosen value then solve for the value of the other variable. There are an infinite number of correct solutions for a given equation in two variables.

For the first solution, we will choose x to be 0.	For the second solution, we will choose x to be 2.	For the third solution, we will choose y to be -3.
$3x + 2y = 6$	$3x + 2y = 6$	$3x + 2y = 6$
$3(\quad) + 2y = 6$	$3(\quad) + 2y = 6$	$3x + 2(\quad) = 6$
$\underline{\qquad} = 6$	$\underline{\qquad} = 6$	$\underline{\qquad} = 6$
$\underline{\qquad} = \underline{\qquad}$	$\underline{\qquad} = \underline{\qquad}$	$\underline{\qquad} = \underline{\qquad}$
	$\underline{\qquad} = \underline{\qquad}$	$\underline{\qquad} = \underline{\qquad}$
Solution: (,)	Solution: (,)	Solution: (,)

    We can summarize the solutions in a table.

x	y	**Ordered Pair**

NOTES

Name: Date:
Instructor: Section:

● **PRACTICE PROBLEMS**

Determine whether the given ordered pair is a solution for the equation.

1. $(-3,-5); 5x+3y=15$ 1._____

2. $(2,6); y=3x+11$ 2._____

3. $\left(0,\dfrac{5}{6}\right); 4x+6y=5$ 3._____

Find three solutions for the given equation. Then graph. (Answers may vary for the three solutions.).

● 4. $y=-3x+2$ 4.

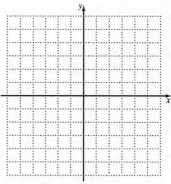

5. $y=4x$ 5.

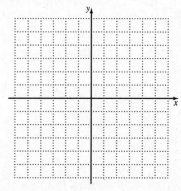

●

79

6. $y = -\dfrac{1}{4}x + 3$

6.

7. $4x - 5y = 20$

7.

Solve.

8. A businesswoman buys a new computer for $1600. For each year that it is in use, she can deduct its depreciated value. The equation $c = -200n + 1600$ gives the value after n years of use.

    a. Find the value of the computer after 2 years.

    b. In how many years will the computer be worth half of its initial value?

    c. After how many years will the computer be worth $0?

    d. Graph the equation with n along the horizontal axis and c along the vertical axis. Because n and c are nonnegative, the graph is restricted to the first quadrant only.

8a. _____

b. _____

c. _____

d.

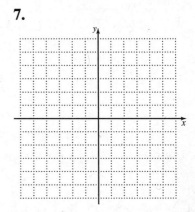

Name: Date:
Instructor: Section:

Chapter 4   GRAPHING LINEAR EQUATIONS AND INEQUALITIES

4.3 Graphing Using Intercepts

KEY VOCABULARY

Term	Definition	Example
x-intercept		
y-intercept		

KEY PROPERTIES, PROCEDURES, OR STRATEGIES

Finding the *x*- and *y*-intercepts

To find an *x*-intercept:	To find a *y*-intercept:

Intercepts for $y = mx$

In the Language of Math	In Your Own Words

The *y*-intercept for $y = mx + b$

In the Language of Math	In Your Own Words

Intercepts for $y = c$

In the Language of Math	In Your Own Words

Intercepts for $x = c$

In the Language of Math	In Your Own Words

GUIDED EXAMPLE

Find the x- and y-intercepts for $4x + 2y = 8$.

Solution

For the x-intercept, replace y with 0 and solve for x.

$$4x + 2y = 8$$

$$4x + 2(\underline{\quad}) = 8$$

$$\underline{\qquad} = 8$$

$$\underline{\qquad} = \underline{\qquad}$$

x-intercept: $(\quad , \quad)$

For the y-intercept, replace x with 0 and solve for y.

$$4x + 2y = 8$$

$$4(\underline{\quad}) + 2y = 8$$

$$\underline{\qquad} = 8$$

$$\underline{\qquad} = \underline{\qquad}$$

y-intercept: $(\quad , \quad)$

NOTES

PRACTICE PROBLEMS

Find the x- and y-intercepts.

1. $6x + 5y = 30$ 1._____

2. $y = 2x + 5$ 2._____

3. $x - 4 = 0$ 3._____

Graph using the x- and y-intercepts.

4. $x - 5y = 10$ **4.**

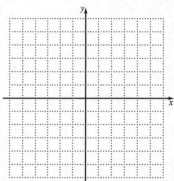

5. $y = 2x$ **5.**

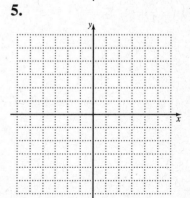

6. $2x - 2 = y$

6.

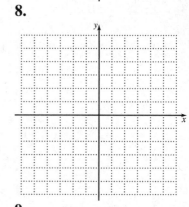

7. $2x + y = 4$

7.

8. $y + 3 = 0$

8.

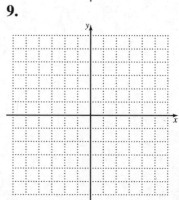

9. $x - 1 = 0$

9.

Chapter 4   GRAPHING LINEAR EQUATIONS AND INEQUALITIES

4.4 Slope-Intercept Form

KEY VOCABULARY

Term	Definition	Example
Slope		

KEY PROPERTIES, PROCEDURES, OR STRATEGIES

Graphs of $y = mx$

Graphing Equations in Slope-Intercept Form

Equation of a Line Given Its Slope and y-Intercept

The Slope Formula

Zero Slope

In the Language of Math	In Your Own Words

Undefined Slope

In the Language of Math	In Your Own Words

GUIDED EXAMPLE

Find the slope of the line connecting the given points.

$(2,9)$ and $(-4,6)$

Solution

Using $m = \dfrac{y_2 - y_1}{x_2 - x_1}$, replace the variables with their corresponding values and then simplify. Let $(2,9)$ be (x_1, y_1) and $(-4,6)$ be (x_2, y_2).

$$m = $$

PRACTICE PROBLEMS

Graph each set of equations on the same grid. For each set of equations, compare the slopes, y-intercepts, and their effects on the graphs.

1. $y = \dfrac{1}{4}x$

       $y = x$

       $y = 4x$

1.

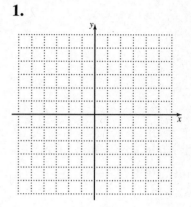

2. $y = -\dfrac{2}{3}x$

       $y = -x$

       $y = -\dfrac{3}{2}x$

2.

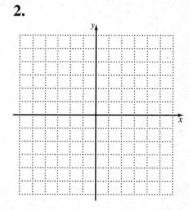

Determine the slope and the y-intercept. Then graph the equation.

3. $2x + y = 4$

3. _____

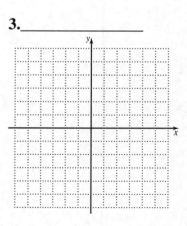

87

4. $2x + 5y = 10$

4._____

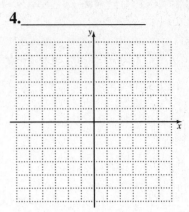

Write the equation of the line in slope-intercept form given the slope and the coordinates of the y-intercept.

5. $m = 4; \left(0, -\dfrac{1}{2}\right)$

5._____

6. $m = -0.4; (0, 2.3)$

6._____

Find the slope of the line through the given points.

7. $(2, 3), (5, 7)$

7._____

8. $(-4, 2), (3, 5)$

8._____

9. $(1, -7), (8, 0)$

9._____

10. $(5, 4), (5, -9)$

10._____

Chapter 4   GRAPHING LINEAR EQUATIONS AND INEQUALITIES

4.5 Point-Slope Form

KEY PROPERTIES, PROCEDURES, OR STRATEGIES

Using the Point-Slope Form of the Equation of a Line

Parallel Lines

In the Language of Math	In Your Own Words

Perpendicular Lines

In the Language of Math	In Your Own Words

NOTES

Name: Date:

Instructor: Section:

GUIDED EXAMPLES

1. Write the equation of the line passing through the points $(4,5)$ and $(12,6)$. Write the equation in slope-intercept form.

   Solution

   First calculate the slope. Use $m = \dfrac{y_2 - y_1}{x_2 - x_1}$.

   $m = $ _____

   Now use the point slope form, then isolate y to write the slope-intercept form. Use either point in the point-slope equation.

   $$y - y_1 = m(x - x_1)$$

   $y - \boxed{} = \boxed{}\left(x - \boxed{}\right)$

   _____ **Simplify.**

   _____ **Isolate y.**

2. Write the equation of the line that passes through $(6,-6)$ and is parallel to $6x - 5y = 2$. Write the equation in slope-intercept form.

   Solution

   First find the slope of the given line $6x - 5y = 2$. Slope: $\boxed{}$

   Therefore the slope of the parallel line will be $\boxed{}$

   Use the point-slope form to write the equation of the parallel line.

   $$y - y_1 = m(x - x_1)$$

   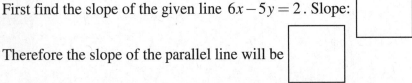

   $y - \boxed{} = \boxed{}\left(x - \boxed{}\right)$

   _____ **Simplify.**

   _____ **Isolate y.**

90

PRACTICE PROBLEMS

Write the equation of the line in slope-intercept form with the given slope passing through the given point.

1. $m = -8; (7,9)$ 1._____

2. $m = \dfrac{3}{2}; (2,-4)$ 2._____

Write the equation of the line passing through the given points. Write the equation in slope-intercept form.

3. $(-5,-8), (-9,-6)$ 3._____

4. $(2,4), (6,5)$ 4._____

Write the equation of the line in slope-intercept form given the y-intercept and one other point.

5. $(4,0), (0,5)$ 5._____

6. $(2,-1), (0,-5)$ 6._____

Write the equation of the line through the given points in the form Ax + By = C, where A, B, and C are integers and A > 0.

7. $(5,3), (1,10)$ 7._____

8. $(4,-8), (-8,7)$ **8.**_____

Write the equation of the line that passes through the given point and is parallel to the given line.
a.   Write the equation in slope-intercept form.
b.   Write the equation in the form Ax + By = C, where A, B, and C are integers and A > 0.

9. $(0,2); y = 5x - 8$ **9a.**_____

                                                  b._____

10. $(2,-2); 4x - 7y = 6$ **10a.**_____

                                                  b._____

Write the equation of the line that passes through the given point and is perpendicular to the given line.
a.   Write the equation in slope-intercept form.
b.   Write the equation in the form Ax + By = C, where A, B, and C are integers and A > 0.

11. $(2,10); y = \dfrac{1}{2}x + 5$ **11a.**_____

                                                  b._____

12. $(2,9); y = -5x + 3$ **12a.**_____

                                                  b._____

Determine whether the given lines are parallel, perpendicular, or neither.

13. $y = 7x + 2$ **13.**_____
       $y = 7x - 2$

14. $x = 3$ **14.**_____
       $y = -9$

Name: Date:

Instructor: Section:

Chapter 4   GRAPHING LINEAR EQUATIONS AND INEQUALITIES

4.6 Graphing Linear Inequalities

KEY PROPERTIES, PROCEDURES, OR STRATEGIES

Checking an Ordered Pair

Graphing Linear Inequalities

NOTES

Name: Date:
Instructor: Section:

GUIDED EXAMPLE
Graph the linear inequality. $y > 5x - 4$

Solution
First, graph the related equation $y = 5x - 4$. Two ordered pairs that satisfy $y = 5x - 4$

are (,   ) and (,   ).

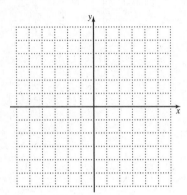

Because the inequality is _____,

we draw a [solid / dashed] line to indicate

that ordered pairs on this boundary line

[are / are not] solutions.

Now choose an ordered pair on one side of the line and test this ordered pair in the inequality. Choose any point that does not lie on the boundary line.

$y > 5x - 4$

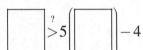

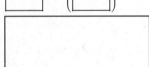

If the statement is false, the ordered pair is *not* a solution for the inequality. If the statement is true, the ordered pair *is* a solution for the inequality.

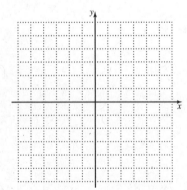

If the ordered pair satisfies the inequality,

shade the region that contains it. If the

ordered pair does not satisfy the inequality,

shade the region on the other side of the

boundary line.

Confirm that the shading is correct by choosing an ordered pair in the shaded region. Check that this ordered pair satisfies the inequality.

94

Name: Date:

Instructor: Section:

PRACTICE PROBLEMS

Determine whether the ordered pair is a solution for the linear inequality.

1. $(-5,8); y < -7x - 2$

1._____

2. $(-9,2); 5x + y \le -6$

2._____

3. $(-2,9); 2x + y > 5$

3._____

Graph the linear inequality.

4. $y \ge -8x + 4$

4.

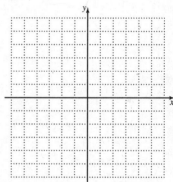

5. $y \le \dfrac{1}{5}x$

5.

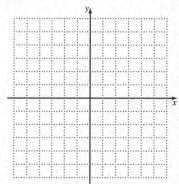

6. $3x + y < 1$

6.

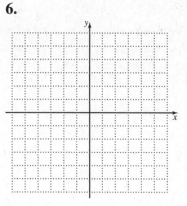

7. $y + 6x > 0$

7.

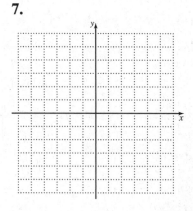

Solve.

8. Jose sells flowers for $6 apiece and vases for $15 apiece. Write an inequality that shows the possible combinations of x flowers and y vases that Jose needs to sell in order to earn at least $120. Then graph the inequality.

8. _____

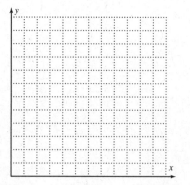

Chapter 4   GRAPHING LINEAR EQUATIONS AND INEQUALITIES

4.7 Introduction to Functions and Function Notation

KEY VOCABULARY

Term	Definition	Example
Relation		
Domain		
Range		
Function		

KEY PROPERTIES, PROCEDURES, OR STRATEGIES

The Vertical Line Test

Finding the Value of a Function

Name: Date:

Instructor: Section:

GUIDED EXAMPLES

For each graph, determine the domain and range. Then state whether the relation is a function.

a)

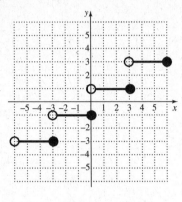

What are all of the *x*-values that have a corresponding *y*-value?

Domain:

What are all of the *y*-values that have a corresponding *x*-value?

Range:

To determine whether the relation is a function, perform a vertical line test. If the relation is not a function, show a vertical line that intersects the graph at two or more points.

Is the relation a function?

b)

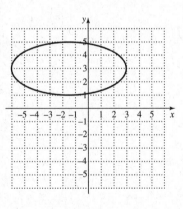

What are all of the *x*-values that have a corresponding *y*-value?

Domain:

What are all of the *y*-values that have a corresponding *x*-value?

Range:

To determine whether the relation is a function, perform a vertical line test. If the relation is not a function, show a vertical line that intersects the graph at two or more points.

Is the relation a function?

98

PRACTICE PROBLEMS

Determine the domain and range of the relation.

1. $\{(7,2),\ (24,-9),\ (38,2),\ (7,8),\ (50,8)\}$

    1._____

    _____

2. $\{(6,4),\ (16,-2),\ (35,4),\ (6,2),\ (59,2)\}$

    2._____

    _____

Determine whether the relation is a function and explain your answer.

3. $\{(5,4),\ (2,4),\ (5,3)\}$

    3._____

    _____

    _____

4. $\{(7,5),\ (-2,-4),\ (2,-1),\ (3,-7)\}$

    4._____

    _____

    _____

Find the value of the function.

5. $f(x)=x^2-x+10$

    a. $f(1)$

    b. $f(-5)$

    c. $f(0)$

    5a._____

    b._____

    c._____

6. $f(x) = \sqrt{x+7}$

    a. $f(0)$

    b. $f(-10)$

    c. $f(b)$

6a._____

b._____

c._____

Use the given graph to find the value of the function.

7. $f(-3)$

7._____

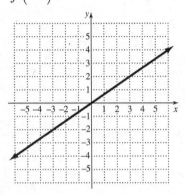

Graph.

8. $f(x) = \dfrac{4}{5}x + 3$

8.

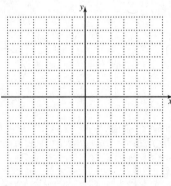

9. $f(x) = -\dfrac{3}{2}x - 2$

9.

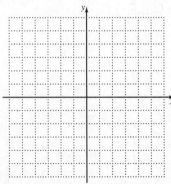

Chapter 5   POLYNOMIALS

5.1 Exponents and Scientific Notation

KEY VOCABULARY

Term	Definition	Example
Scientific notation		

KEY PROPERTIES, PROCEDURES, OR STRATEGIES

Evaluating Exponential Forms with Negative Bases

Raising a Quotient to a Power

In the Language of Math	In Your Own Words

Nonpositive Integer Exponent Rules

In the Language of Math	In Your Own Words

Changing Scientific Notation (Positive Exponent) to Standard Form

Changing Scientific Notation (Negative Exponent) to Standard Form

Changing Standard Form to Scientific Notation

GUIDED EXAMPLES

Write each number in standard form.

    a) 2.13×10^4

Solution

Multiplying 2.13 by 10^4 means that the decimal point will move _____ places to

the [left / right].

$2.13 \times 10^4 =$

    b) 4.57×10^{-8}

Solution

Multiplying 4.57 by 10^{-8} means that the decimal point will move _____ places to

the [left / right].

$4.57 \times 10^{-8} =$

PRACTICE PROBLEMS

Evaluate the exponential form.

1. 9^0

1._____

2. -7^2

2._____

3. $\left(\dfrac{2}{5}\right)^3$

3._____

4. $(-4)^3$

4._____

Rewrite each expression with positive exponents; then if the expression is numeric, evaluate it.

5. 6^{-2}

5._____

6. $\left(\dfrac{4}{5}\right)^{-4}$

6._____

7. $(-3)^{-3}$

7._____

Write the number in standard form.

8. 4.12×10^5 **8.**_____

9. 7.32×10^6 **9.**_____

10. 6.11×10^{-5} **10.**_____

11. 1.43×10^{-7} **11.**_____

Write the number in scientific notation.

12. Sales totals for the Automation Company were   **12.**_____
$25,000,000.

13. It is estimated that in the year 2020 the population   **13.**_____
of the world will be 7,518,000,000.

14. One cubic orc is approximately equal to   **14.**_____
0.0000000048 cubic balrogs.

Chapter 5   POLYNOMIALS

5.2 Introduction to Polynomials

KEY VOCABULARY

Term	Definition	Example
Monomial		
Coefficient of a monomial		
Degree of a monomial		
Polynomial		
Polynomial in one variable		
Binomial		
Trinomial		
Degree of a polynomial		

KEY PROPERTIES, PROCEDURES, OR STRATEGIES

Writing a Polynomial in Descending Order of Degree

GUIDED EXAMPLE

Combine like terms and write the resulting polynomial in descending order of degree.

$$11r - 8r^2 + 22d^3 + 9r - 5d^3 + 88r^2 + 16d^3$$

Solution

$$11r - 8r^2 + 22d^3 + 9r - 5d^3 + 88r^2 + 16d^3$$

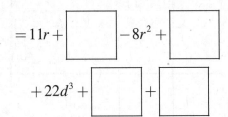   **Collect like terms.**

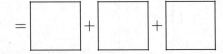   **Combine like terms.**

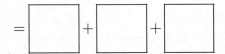   **Rearrange the terms so that the highest degree term is first, then the next highest degree, and so on.**

NOTES

PRACTICE PROBLEMS

Determine whether the expression is a monomial.

1. $-6x^2y^4$

1._____

2. $-8x^{-3}y^2$

2._____

Identify the coefficient and degree of each monomial.

3. $6t^7$

3._____

4. mn^2

4._____

Indicate whether the expression is a monomial, binomial, or trinomial or has no special polynomial name. If the expression is a polynomial, give the degree.

5. $x^2 - 2x + 1$

5._____

6. 68

6._____

7. $3p^3m^4 - 5p^2m^3 + 5p^3m - 5pm^3 + 4p^4m^3$

7._____

Identify the degree of each polynomial.

8. $x^5 - 6x + x^8 - 5x^6$

8._____

9. $1.2x^5 - 7.3x^3 + 4.5x + 9.5$

9._____

Name: Date:

Instructor: Section:

Evaluate the polynomial using the given values.

10. $3x^2 - 3x + 1; x = 4$

10._____

11. $a^3 - 3a^2 + 4a + 7; a = 5$

11._____

Evaluate.

12. A 12-ounce beverage can has a height of 7.3 inches and a radius of 1.6 inches. The polynomial $2\pi rh + 2\pi r^2$ describes the surface area of a cylindrical can. Evaluate the polynomial for $h = 7.3$ inches and $r = 1.6$ inches to find the surface area of the can. Round the result to the nearest tenth.

12._____

Write the polynomial in descending order of degree.

13. $x^5 + x + 8x^3 + 7 + 6x^2$

13._____

Combine like terms and write the resulting polynomial in descending order of degree.

14. $9b^5 + b^2 - b^3 - 6b^5 - 7b^2$

14._____

15. $11a^2 p + 4p^2 - 33a^2 p - 6p^2$

15._____

Name: Date:
Instructor: Section:

Chapter 5   POLYNOMIALS

5.3 Adding and Subtracting Polynomials

KEY PROPERTIES, PROCEDURES, OR STRATEGIES

Adding Polynomials

Subtracting Polynomials

GUIDED EXAMPLE

1. Add and write the resulting polynomial in descending order of degree.

$$\left(3y^4 - 5y^3 + 2y - 7\right) + \left(y^4 + 2y^3 - y^2 + 6y + 1\right)$$

Solution

Combine like terms. Combining in order of degree places the resulting polynomial in descending order of degree.

$$\left(3y^4 - 5y^3 + 2y - 7\right) + \left(y^4 + 2y^3 - y^2 + 6y + 1\right)$$

$$=$$

NOTES

GUIDED EXAMPLES

2. Write an expression in simplest form for the perimeter of the triangle.

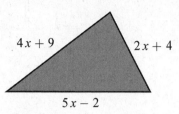

Solution

Understand Perimeter means the total distance around the shape. We need to add the lengths of all the sides of the shape.

Plan   The lengths of the sides are represented by polynomials. Add the polynomials to represent the perimeter.

Execute   Perimeter   =   Side   +   Side   +   Side

           =   +   +

           =

           =

Answer   The expression for the perimeter is

Check   Choose a value for x and evaluate the original expressions for the side lengths, determine the corresponding numeric perimeter, and evaluate the perimeter expression using the same value for x to verify that we get the same numeric perimeter.

3. Subtract and write the resulting polynomial in descending order of degree.
$$\left(-11m^4 - 27m^3 + 6m - 8\right) - \left(7m^4 - 3m^3 + 9m + 1\right)$$

Solution

Write an equivalent addition statement.

$$\left(-11m^4 - 27m^3 + 6m - 8\right) - \left(7m^4 - 3m^3 + 9m + 1\right)$$
$$\downarrow$$

=   **Change the minus sign between the polynomials in parentheses to a plus sign, and change all signs in the subtrahend.**

=   **Combine like terms.**

PRACTICE PROBLEMS

Add and write the resulting polynomial in descending order of degree.

1. $(7x+8)+(-8x+6)$

1._____

2. $(5x+9y)+(-6x+8y)$

2._____

3. $(-9x+4)+(x^2+x-6)$

3._____

4. $(9x^2-6x+15)+(7x^2+8x-89)$

4._____

5. $(2+8x+3x^2+7x^3)+(9-8x+3x^2-7x^3)$

5._____

6. $(-2x^4-3x^3-3x^2+x-14)+(-3x^4+9x^3+x^2+14x+7)$

6._____

Write an expression for the perimeter in simplest form.

7. A rectangle with width $6x-8$ and length $3x+9$

7._____

111

Subtract and write the resulting polynomial in descending order of degree.

8. $(4x+5)-(2x+9)$

8._____

9. $\left(-3v^2+9v+7\right)-\left(7v^2+11v-2\right)$

9._____

10. $\left(-4a^3-3a^2+6a+9\right)-\left(-4a^3+9a^2+10a-4\right)$

10._____

11. $\left(-y^5+3y^4+6y^2-y-15\right)-\left(-3y^5+y^2-y-5\right)$

11._____

12. $\left(4w^2+9wt-2t^2\right)-\left(6w^2-7wt+16t^2\right)$

12._____

Chapter 5   POLYNOMIALS

5.4 Exponent Rules and Multiplying Monomials

KEY PROPERTIES, PROCEDURES, OR STRATEGIES

Product Rule for Exponents

In the Language of Math	In Your Own Words

Multiplying Monomials

A Power Raised to a Power

In the Language of Math	In Your Own Words

Raising a Product to a Power

In the Language of Math	In Your Own Words

Simplifying a Monomial Raised to a Power

```




```

GUIDED EXAMPLES

1. Multiply.

    $\left(5n^3\right)\left(8n^6\right)$

    Solution

    $\left(5n^3\right)\left(8n^6\right)$

    = [   ]   **Multiply the coefficients and add the exponents of the like bases.**

    = [   ]   **Simplify.**

2. Simplify.

    $\left(2x^3z^4\right)^6$

    Solution

    $\left(2x^3z^4\right)^6$

    = [   ]   **Write the coefficient, 2, raised to the 6th power and multiply the exponents on the variables by 6.**

    = [   ]   **Simplify.**

NOTES

PRACTICE PROBLEMS

Multiply.

1. $v^3 \cdot v^6$

1._____

2. $7^9 \cdot 7^3$

2._____

3. $7r^2 \cdot 5r$

3._____

4. $\left(-5x^7 y^7\right)\left(4x^6 y^2\right)$

4._____

5. $\left(3qr\right)\left(6q^2 r^4\right)\left(2q^5\right)$

5._____

6. $\left(5xy\right)\left(4x^3 y^4\right)\left(3x^2\right)$

6._____

Write an expression in simplest form for the area of the figure.

7.

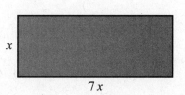

x

$7x$

7._____

115

Multiply and write your answer in scientific notation.

8. $\left(3.1\times10^{6}\right)\left(2.4\times10^{2}\right)$ **8.**_____

9. $\left(7.56\times10^{7}\right)\left(3.48\times10^{-4}\right)$ **9.**_____

Simplify.

10. $\left(c^{2}\right)^{8}$ **10.**_____

11. $\left(-x^{7}\right)^{4}$ **11.**_____

12. $\left(\dfrac{1}{3}x^{2}y\right)^{3}$ **12.**_____

13. $\left(-r^{4}s\right)^{3}\left(-r^{5}s^{4}\right)^{2}$ **13.**_____

14. $\left(3a\right)^{2}\left(a^{3}b\right)\left(-5ab\right)^{2}$ **14.**_____

Chapter 5   POLYNOMIALS

5.5 Multiplying Polynomials; Special Products

KEY VOCABULARY

Term	Definition	Example
Conjugates		

KEY PROPERTIES, PROCEDURES, OR STRATEGIES

Multiplying a Polynomial by a Monomial

Multiplying Polynomials

Multiplying Conjugates

In the Language of Math	In Your Own Words

Squaring a Binomial

In the Language of Math	In Your Own Words

117

GUIDED EXAMPLES

Multiply.

   a) $5x\left(6x^2 - 8x + 3\right)$

   Solution

   $5x\left(6x^2 - 8x + 3\right)$

   $= 5x \cdot \boxed{} + 5x \cdot \boxed{} + 5x \cdot \boxed{}$ **Multiply each term in the polynomial by 5x.**

   $= \boxed{}$

   b) $\left(8x + 1\right)\left(6x - 4\right)$

   Solution

   $\left(8x + 1\right)\left(6x - 4\right)$

       First   Outer   Inner   Last

   $= \boxed{} + \boxed{} + \boxed{} + \boxed{}$ **Multiply each term in 8x + 1 by each term in 6x – 4. (Think FOIL.)**

   $= \boxed{}$

   $= \boxed{}$ **Combine like terms.**

   $= \boxed{}$

NOTES

PRACTICE PROBLEMS

Multiply the polynomial by the monomial.

1. $-5x(x-3)$

1._____

2. $4x(4x^2-9x+9)$

2._____

3. $-4y^4(9y^2+4y-6)$

3._____

Multiply the binomials. (Use FOIL.)

4. $(b+11)(b-9)$

4._____

5. $(3x-1)(4x+3)$

5._____

6. $(4v-5f)(4v+4f)$

6._____

Multiply the polynomials.

7. $(x^2+x+7)(x-7)$

7._____

8. $(9x+5)(2x^2+2x+4)$

8._____

119

9. $\left(r^2+r-1\right)\left(r^2+7r-9\right)$

9._____

State the conjugate of the given binomial.

10. $a+7$

10._____

11. $\dfrac{1}{2}b-2c$

11._____

Multiply using the rules for special products.

12. $(s+6)(s-6)$

12._____

13. $(3r+5)(3r-5)$

13._____

14. $(y-5)^2$

14._____

15. $(5x+4y)^2$

15._____

Chapter 5   POLYNOMIALS

5.6 Exponent Rules and Dividing Polynomials

KEY PROPERTIES, PROCEDURES, OR STRATEGIES

Dividing Monomials

Dividing a Polynomial by a Monomial

In the Language of Math	In Your Own Words

Dividing a Polynomial by a Polynomial

Exponents Summary

Name: Date:

Instructor: Section:

GUIDED EXAMPLES

1. Divide and write the result with a positive exponent.

$$\frac{g^{-6}}{g^{-3}}$$

Solution

$\dfrac{g^{-6}}{g^{-3}} = $ [   ]   **Subtract the exponents and keep the same base.**

$= $ [   ]   **Rewrite the subtraction as addition.**

$= $ [   ]   **Simplify.**

$= $ [   ]   **Write with a positive exponent.**

2. Simplify. Write the answer with a positive exponent.

$$\frac{\left(x^3\right)^{-2}}{x^6 \cdot x^{-8}}$$

Solution

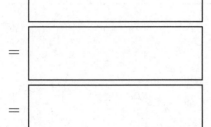

$\dfrac{\left(x^3\right)^{-2}}{x^6 \cdot x^{-8}} = $ [   ]   **In the numerator, use the rule for raising a power to a power. In the denominator, use the product rule for exponents.**

$= $ [   ]

$= $ [   ]   **Use the quotient rule for exponents.**

$= $ [   ]   **Write the subtraction as an equivalent addition.**

$= $ [   ]

$= $ [   ]   **Write with a positive exponent.**

122

PRACTICE PROBLEMS

Simplify using the rules of exponents. Write all answers with positive exponents.

1. $\dfrac{3^8}{3^5}$

1._____

2. $\dfrac{x^3}{x^{-2}}$

2._____

Divide and write your answers in scientific notation.

3. $\dfrac{-18.2 \times 10^{-7}}{2.6 \times 10^{-4}}$

3._____

4. $\dfrac{-11.6 \times 10^{-4}}{2.9 \times 10^{3}}$

4._____

Divide the monomials.

5. $\dfrac{15a^4b^5}{-3ab}$

5._____

6. $\dfrac{-2st^3}{10s^4t}$

6._____

123

Divide the polynomial by the monomial.

7. $\dfrac{7x+28y}{7}$

7._____

8. $\dfrac{9a^5-9a^3+9}{3a}$

8._____

Use long division to divide the polynomials.

9. $\dfrac{x^2+7x+6}{x+6}$

9._____

10. $\dfrac{x^2-x-46}{x-7}$

10._____

11. $\dfrac{x^3+729}{x+9}$

11._____

12. $\dfrac{12y^2-31y+147}{3y+8}$

12._____

Simplify. Write all answers with positive exponents.

13. $\left(\dfrac{4}{5}\right)^{-3}$

13._____

14. $\left(\dfrac{1}{6}\right)^{-1}$

14._____

Chapter 6   FACTORING

6.1 Greatest Common Factor and Factoring by Grouping

KEY VOCABULARY

Term	Definition	Example
Factored form		
Greatest common factor (GCF)		

KEY PROPERTIES, PROCEDURES, OR STRATEGIES

Listing Method for Finding GCF

Prime Factorization Method for Finding GCF

125

Factoring a Monomial GCF Out of a Polynomial

Factoring by Grouping

GUIDED EXAMPLE

Factor $24xy^3 + 32y^2$.

Solution

First find the GCF of $24xy^3$ and $32y^2$. Write the prime factorization of each monomial, treating the variables like prime factors.

$24xy^3 = $ [ ] $32y^2 = $ [ ]

GCF: [ ]

$24xy^3 + 32y^2 = $ [ ]   **Write the polynomial as the product of the GCF and the quotient of the polynomial and the GCF.**

$= $ [ ]   **Separate the terms.**

$= $ [ ]   **Divide the terms by the GCF.**

126

PRACTICE PROBLEMS

List all natural number factors of the given number.

1.   18

1._____

2.   625

2._____

Find the GCF.

3.   40, 100

3._____

4.   32x, 20

4._____

Factor by factoring out the GCF.

5. $x^2 - 5x$

5._____

6. $5a^2 y - 15ay$

6._____

7. $8m^3 n^3 + 48m^3 n^2$

7._____

8. $3x^7 y^5 + 21x^5 y^4 + 12xy$

8._____

Factor by factoring out the negative of the GCF.

9. $-6z^2 + 14z$ **9.**_____

Factor out the polynomial GCF.

10. $x(c-4) + t(c-4)$ **10.**_____

11. $6m(7m-4) - 5(7m-4)$ **11.**_____

Factor by grouping.

12. $sq + sz + fq + fz$ **12.**_____

13. $s^2 + 3s + 7s + 21$ **13.**_____

14. $4z^2 + 8z - az - 2a$ **14.**_____

15. $r^2 - 10tw + 2wr - 5tr$ **15.**_____

Chapter 6   FACTORING

6.2 Factoring Trinomials of the Form $x^2 + bx + c$

KEY PROPERTIES, PROCEDURES, OR STRATEGIES

Factoring $x^2 + bx + c$

GUIDED EXAMPLE

1. Factor $t^2 + t - 12$.

Solution
We must find a pair of numbers whose product is -12 and whose sum is 1. Because the product is negative, the two numbers must have different signs. Because the sum is positive, the number with the greater absolute value will be positive.

Product	Sum

Once we have found the correct combination, write the polynomial as the product of two binomials, where the numbers we found are the second terms in the binomials.

$$t^2 + t - 12 = \boxed{}$$

NOTES

GUIDED EXAMPLES

2. Factor $b^2 + 6bh - 16h^2$.

Solution

The variable h is in the last term, so think of the middle term $6bh$ as $6hb$ with "coefficient" $6h$. We must find a pair of terms whose product is $-16h^2$ and whose sum is $6h$.

Product	Sum

$$b^2 + 6bh - 16h^2 = \boxed{}$$

3. Factor $7z^6 - 28z^5 - 147z^4$.

Solution

Whenever factoring polynomials, the first step should be to look for a monomial GCF among the terms.

GCF: $\boxed{}$

Now try to factor the trinomial to two binomials. We are looking for two numbers whose product is negative and whose sum is negative, so the numbers will have different signs and the number with the greater absolute value will be negative.

Product	Sum

$$7z^6 - 28z^5 - 147z^4 = \boxed{}$$

Name: Date:

Instructor: Section:

PRACTICE PROBLEMS

Fill in the missing values in the factors.

1. $x^2 + 15x + 44 = (x+4)(x+\underline{})$

    1._____

2. $x^2 - 12x + 35 = (x-5)(x-\underline{})$

    2._____

3. $x^2 - 6x - 55 = (x+5)(x-\underline{})$

    3._____

4. $x^2 + 9x - 10 = (x-1)(x+\underline{})$

    4._____

Factor. If the polynomial is prime, so state.

5. $t^2 + 6t + 8$

    5._____

6. $b^2 - 14b + 45$

    6._____

7. $r^2 - 17r + 70$

    7._____

8. $w^2 - w - 30$

    8._____

131

9. $s^2 + 4s + 77$ 9._____

Factor the trinomials containing two variables. If the polynomial is prime, so state.

10. $w^2 + 13wz + 42z^2$ 10._____

11. $b^2 - 6bx - 27x^2$ 11._____

12. $s^2 - 8sz + 15z^2$ 12._____

Factor completely.

13. $2x^3 + 14x^2 + 20x$ 13._____

14. $4x^7 - 8x^6 - 32x^5$ 14._____

15. $6c^2 - 60c + 126$ 15._____

Chapter 6   FACTORING

6.3 Factoring Trinomials of the Form $ax^2 + bx + c$, where $a \neq 1$

KEY PROPERTIES, PROCEDURES, OR STRATEGIES

Factoring by Trial and Error

Factoring $ax^2 + bx + c$, where $a \neq 1$, by Grouping

GUIDED EXAMPLE

Factor $32v^2 + 20v - 3$ by grouping.

Solution

For the trinomial $32v^2 + 20v - 3$,

$a = $ _____, $b = $ _____, $c = $ _____

Find the product ac: _____

Now find two factors of this product whose sum is b. Because the product is negative, the factors have different signs. Because the sum is positive, the factor with the greater absolute value must be positive.

Factors of ac	Sum of Factors of ac

Now use the factors found in the table to write the middle term of the polynomial as a sum of two like terms.

$32v^2 + 20v - 3 = $

Factor the new polynomial by grouping.

Answer: $32v^2 + 20v - 3 = $

PRACTICE PROBLEMS

Factor completely. If prime, so state.

1. $t^2 - 12t + 35$ 1._____

2. $t^2 - 5t + 6$ 2._____

3. $b^2 + 5b + 66$ 3._____

4. $4a^2 + 21a + 5$ 4._____

5. $w^2 - 9wz + 20z^2$ 5._____

6. $a^2 - 16af + 63f^2$ 6._____

7. $4s^2 - 22s + 10$ 7._____

Factor by grouping. If prime, so state.

8. $\quad 4u^2 + 9u + 5$

8._____

9. $\quad 18v^2 - 85v + 18$

9._____

10. $56u^2 - 122u + 42$

10._____

11. $28b^2 + 17b - 3$

11._____

12. $25c^2 + 60c + 27$

12._____

13. $9a^2 + 18a + 8$

13._____

14. $30c^2 - 11cd - 30d^2$

14._____

15. $30u^3 + 145u^2 - 210u$

15._____

Chapter 6   FACTORING

6.4 Factoring Special Products

KEY PROPERTIES, PROCEDURES, OR STRATEGIES

Factoring Perfect Square Trinomials

Factoring a Difference of Squares

Factoring a Difference of Cubes

Factoring a Sum of Cubes

NOTES

GUIDED EXAMPLE

Factor $49v^2 - 56v + 16$.

Solution

Look at the terms in the trinomial $49v^2 - 56v + 16$ to determine whether the trinomial fits the form of a perfect square.

$$49v^2 = \left(\quad\right)^2$$

$$16 = \left(\quad\right)^2$$

$$56v = 2\cdot\left(\quad\right)\cdot\left(\quad\right)$$

The first and last terms of the trinomial are perfect squares and the middle term is equal to twice the product of the square roots of the first and last terms. So $49v^2 - 56v + 16$ is a perfect square trinomial fitting the form $a^2 - 2ab + b^2$. We can write the factored form as $(a-b)^2$, where

$$a = \underline{\qquad} \text{ and } b = \underline{\qquad}$$

Answer: $49v^2 - 56v + 16 = \boxed{\qquad\qquad\qquad}$

NOTES

PRACTICE PROBLEMS

Factor the trinomials that are perfect squares. If the trinomial is not a perfect square, write *not a perfect square.*

1. $s^2 + 12s + 36$

   1._____

2. $v^2 - 16v + 64$

   2._____

3. $25a^2 - 90a + 81$

   3._____

4. $9c^2 - 24cg + 16g^2$

   4._____

5. $25r^2 + 70rt + 49t^2$

   5._____

Factor the binomials that are the difference of squares. If prime, so state.

6. $r^2 - 36$

   6._____

7. $4c^2 - 121$

   7._____

8. $9a^2 - 121g^2$

   8._____

139

Name: Date:
Instructor: Section:

Factor the sum or difference of cubes.

9. $r^3 - 27$

9._____

10. $27x^3 - 64b^3$

10._____

11. $b^3 + 64$

11._____

12. $27x^3 + 64g^3$

12._____

Factor. If prime, so state.

13. $25m^2 - \dfrac{36}{25}$

13._____

14. $45a^2 - 245$

14._____

15. $\left(y+z\right)^3 - 8$

15._____

Chapter 6   FACTORING

6.5 Strategies for Factoring

KEY PROPERTIES, PROCEDURES, OR STRATEGIES

Strategies for Factoring

Name: Date:
Instructor: Section:

GUIDED EXAMPLES
Factor.

   a) $a^2 + 13a + 42$
   Solution
   First factor out any monomial GCF. There is no monomial GCF in this polynomial.
   There are three terms in this polynomial, so check to see if it is a perfect square.

   Are the first and third terms both perfect squares? _____

   Now consider the form of the trinomial. It has form _____.

   So find two factors of c whose sum is b and write the factored form as
                 (a + first number)(a + second number).

   Answer: $a^2 + 13a + 42 =$

   b) $ac + ad + bc + bd$
   Solution
   First factor out any monomial GCF. There is no monomial GCF in this polynomial.
   There are four terms in this polynomial, so try factoring by grouping.

   Answer: $ac + ad + bc + bd =$

   c) $54x^3 - 128s^3$
   Solution
   First factor out any monomial GCF. The monomial GCF in this polynomial is _____.
   There are two terms in the binomial factor of this polynomial, so check to see if the
   binomial is a difference of squares, a sum of cubes, or a difference of cubes. If so, use
   the appropriate rule to factor the polynomial completely.

   Answer: $54x^3 - 128s^3 =$

142

PRACTICE PROBLEMS

Factor completely. If prime, so state.

1. $a(c-5)+z(c-5)$ 1._____

2. $24b^2-294$ 2._____

3. v^3+27 3._____

4. $vw+vx+kw+kx$ 4._____

5. $r^2+10r+16$ 5._____

6. t^2+t-30 6._____

7. t^2-t-30 7._____

8. b^2-25 8._____

9. $125x^3 - 216a^3$

9._____

10. $54u^2 - 255u + 54$

10._____

11. $s^4 - 81$

11._____

12. $x^2 - 4x - 45$

12._____

13. $4v^2 - 36v + 81$

13._____

14. $x^2 - 6x$

14._____

15. $9r - 100r^3$

15._____

Chapter 6   FACTORING

6.6 Solving Quadratic Equations by Factoring

KEY VOCABULARY

Term	Definition	Example
Quadratic equation in one variable		

KEY PROPERTIES, PROCEDURES, OR STRATEGIES

Zero-Factor Theorem

In the Language of Math	In Your Own Words

Solving Equations with Two or More Factors Equal to 0

Solving Quadratic Equations Using Factoring

145

Name: Date:
Instructor: Section:

Pythagorean Theorem

```




```

GUIDED EXAMPLE

Solve $r^2 - 30r = -29$.

Solution

$r^2 - 30r = -29$

Write the equation in standard form, $ax^2 + bx + c = 0$.

Factor the variable expression.

Use the zero-factor theorem to solve.

To check, verify that the solutions for the variable satisfy the original equation.

NOTES

Name: Date:
Instructor: Section:

PRACTICE PROBLEMS

Solve using the zero-factor theorem.

1. $(a-25)(a+64)=0$

1._____

2. $(4t+9)(t+8)=0$

2._____

3. $r(r+4)=0$

3._____

Solve the quadratic equations.

4. $(v-5)^2=0$

4._____

5. $x^2-25x+24=0$

5._____

6. $w^2-6w+9=0$

6._____

7. $a^2=-12-7a$

7._____

8. $b^2=16$

8._____

9. $5w^2=26w+24$

9._____

10. $r(r-5)=36$

10._____

Translate to an equation and then solve.

11. The sum of the squares of two consecutive odd positive integers is 74. Find the integers.

11.

12. The length of the top of a rectangular table is 5 meters greater than the width. The area is 104 square meters. Find the dimensions of the table.

12.

13. Use the formula $h = -16t^2 + v_0 t + h_0$, where h is the final height in feet, t is the time of travel in seconds, v_0 is the initial velocity in feet per second, and h_0 is the initial height in feet of an object traveling upward. If an object is thrown upward at 96 feet per second from a height of 4 feet, when will the object be 144 feet off the ground?

13.

Find the length of the hypotenuse.

14.

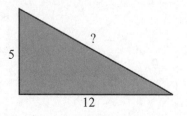

14. _____

Chapter 6   FACTORING

6.7 Graphs of Quadratic Equations and Functions

KEY VOCABULARY

Term	Definition	Example
Quadratic equation in two variables		
Axis of symmetry		
Vertex of a parabola		

KEY PROPERTIES, PROCEDURES, OR STRATEGIES

Graphing Quadratic Equations

Opening of a Parabola

In the Language of Math	In Your Own Words

Graphing Quadratic Functions

GUIDED EXAMPLE

Graph $f(x) = 2x^2 - 8x + 5$.

Solution

Find enough ordered pairs to clearly see the graph. Then connect the points to form the parabola.

x	*y*
0	
1	
2	
3	
4	

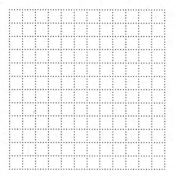

NOTES

Name: Date:
Instructor: Section:

PRACTICE PROBLEMS

Graph.

1. $y = x^2 + 6x - 1$

1.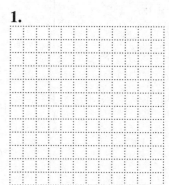

2. $y = x^2 - 9$

2.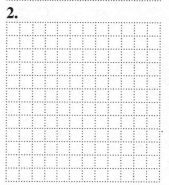

3. $y = 4 - x^2$

3.

4. $y = -x^2 + 2x - 3$

4.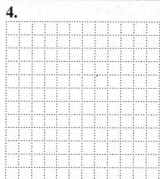

151

5. $f(x) = x^2 - 4x - 1$

5.

6. $f(x) = x^2 - x - 8$

6.

7. $f(x) = -x^2 - 8x - 10$

7.

State whether the graph is the graph of a function. Give the domain and range.

8.

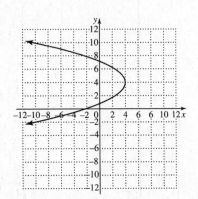

8._____

Chapter 7   RATIONAL EXPRESSIONS AND EQUATIONS

7.1 Simplifying Rational Expressions

KEY VOCABULARY

Term	Definition	Example
Rational expression		

KEY PROPERTIES, PROCEDURES, OR STRATEGIES

Finding Values That Make a Rational Expression Undefined

Simplifying Rational Expressions to Lowest Terms

NOTES

153

Name: _____   Date: _____

Instructor: _____   Section: _____

GUIDED EXAMPLES

1. Find every value for the variable that makes the expression undefined.

$$\frac{x}{x^2+4x-5}$$

Solution

$$x^2+4x-5=0$$ **Set the denominator equal to 0.**

_____   **Factor the polynomial.**

_____   **Use the zero-factor theorem.**

The expression $\dfrac{x}{x^2+4x-5}$ is undefined if x is replaced by [___] or [___].

2. Simplify.

$$\frac{10z-40}{8z-32}$$

Solution

$$\frac{10z-40}{8z-32}=\boxed{}$$ **Factor the numerator and denominator completely.**

$$=\boxed{}$$ **Divide out the common factors.**

$$=\boxed{}$$ **Simplify.**

NOTES

Name: Date:
Instructor: Section:

PRACTICE PROBLEMS

Evaluate the rational expression.

1. $\dfrac{8x-10}{3x}$

    a. when $x=2$
    b. when $x=-3$

    1a._____

    b._____

2. $\dfrac{(-5x)^2}{4x+12}$

    a. when $x=2$
    b. when $x=-3$

    2a._____

    b._____

3. $\dfrac{3x+1}{5x+25}$

    a. when $x=5$
    b. when $x=-5$

    3a._____

    b._____

Find every value for the variable that makes the expression undefined.

4. $\dfrac{4}{x-5}$

    4._____

5. $\dfrac{3x}{x^2-16}$

    5._____

6. $\dfrac{2x+4}{x^2-2x-3}$

    6._____

Name: Date:

Instructor: Section:

Simplify if possible.

7. $\dfrac{1250u^8z^{10}}{20u^4z^2}$ 7._____

8. $\dfrac{15(y-7)}{6(y-7)}$ 8._____

9. $\dfrac{8a-24}{7a-21}$ 9._____

10. $\dfrac{a^3-c^3}{a^2-c^2}$ 10._____

11. $\dfrac{r^2-49}{r^2-14r+49}$ 11._____

12. $\dfrac{w^2+1}{w+1}$ 12._____

13. $\dfrac{ak-av-dk+dv}{ak-av+dk-dv}$ 13._____

14. $\dfrac{6v-12}{2-v}$ 14._____

156

Chapter 7   RATIONAL EXPRESSIONS AND EQUATIONS

7.2 Multiplying and Dividing Rational Expressions

KEY PROPERTIES, PROCEDURES, OR STRATEGIES

Multiplying Rational Expressions

Dividing Rational Expressions

Using Dimensional Analysis to Convert between Units of Measurement

Name: _____ Date: _____

Instructor: _____ Section: _____

GUIDED EXAMPLES

1. Multiply.

$$\frac{125-5z}{49} \cdot \frac{245}{7z-175}$$

Solution

$$\frac{125-5z}{49} \cdot \frac{245}{7z-175} = \boxed{}$$

Factor the numerators and denominators completely.

$$= \boxed{}$$

Divide out the common factors.

$$= \boxed{}$$

Multiply the remaining numerator factors and denominator factors.

2. Divide.

$$\frac{50x^9}{2y^9} \div \frac{625x^4}{10y^3}$$

Solution

$$\frac{50x^9}{2y^9} \div \frac{625x^4}{10y^3} = \boxed{}$$

Write an equivalent multiplication statement by changing the division sign to multiplication and changing the divisor to its reciprocal.

$$= \boxed{}$$

Factor the numerators and denominators completely.

$$= \boxed{}$$

Divide out the common factors.

$$= \boxed{}$$

Multiply the remaining numerator factors and denominator factors.

158

PRACTICE PROBLEMS

Multiply.

1. $\dfrac{11}{m} \cdot \dfrac{m^5}{11}$

    1._____

2. $\dfrac{12y}{7} \cdot \dfrac{y}{3} \cdot \dfrac{2}{y^2}$

    2._____

3. $\dfrac{8-2z}{49} \cdot \dfrac{98}{7z-28}$

    3._____

4. $\dfrac{x^2-7x+10}{x^2-4} \cdot \dfrac{x-2}{x^2-10x+25}$

    4._____

5. $\dfrac{9x}{x^2-4x+4} \cdot \dfrac{x^2-4}{18x^2}$

    5._____

6. $\dfrac{c^2+cf-cd-df}{c^2+5c+cf+5f} \cdot \dfrac{c^2-7c-35+5c}{c^2-7d+cd-7c}$

    6._____

Divide.

7. $\dfrac{3}{r} \div \dfrac{27}{r}$

8. $\dfrac{20x^6}{5y^5} \div \dfrac{16x^4}{10y^2}$

8._____

9. $\dfrac{w^3 - 7w^2 + 12w}{2x} \div \dfrac{2w - 6}{4w + 16}$

9._____

10. $\dfrac{w + 4}{2} \div \dfrac{3w + 12}{8}$

10._____

11. $\dfrac{v^2 + 8v + 16}{v + 3} \div \left(5v^2 + 17v - 12\right)$

11._____

12. $\dfrac{k + 4}{3} \div \dfrac{2k + 8}{9}$

12._____

Use dimensional analysis to convert units of length.

13. 5 yards to inches

13._____

Chapter 7   RATIONAL EXPRESSIONS AND EQUATIONS

7.3 Adding and Subtracting Rational Expressions with the Same Denominator

KEY PROPERTIES, PROCEDURES, OR STRATEGIES

Adding or Subtracting Rational Expressions (Same Denominator)

GUIDED EXAMPLES

1.   Add $\dfrac{c}{c+4} + \dfrac{8}{c+4}$.

Solution

$$\dfrac{c}{c+4} + \dfrac{8}{c+4} = \boxed{}$$ **Add the numerators and keep the same denominator.**

Because c and 8 are not like terms, we express their sum as a polynomial.

2.   Subtract $\dfrac{a^2}{a+2} - \dfrac{4}{a+2}$.

Solution

$$\dfrac{a^2}{a+2} - \dfrac{4}{a+2} = \boxed{}$$ **Subtract the numerators and keep the same denominator.**

$$= \boxed{}$$ **Factor the numerator.**

$$= \boxed{}$$ **Divide out the common factor.**

161

Name: Date:
Instructor: Section:

GUIDED EXAMPLE

3. Add $\dfrac{y^2+8y-8}{3y^3+9y^2-30y}+\dfrac{2y^2+6y+3}{3y^3+9y^2-30y}$.

Solution

$\dfrac{y^2+8y-8}{3y^3+9y^2-30y}+\dfrac{2y^2+6y+3}{3y^3+9y^2-30y}$

= [   ]   Add the numerators
                                   and keep the same
                                   denominator.

= [   ]   Combine like terms.

= [   ]   Factor the numerator
                                   and denominator
                                   completely.

= [   ]   Divide out common
                                   factors.

NOTES

PRACTICE PROBLEMS

Add or subtract. Simplify your answers to lowest terms.

1. $\dfrac{3x}{104} + \dfrac{5x}{104}$

                                                1._____

2. $\dfrac{8}{n^2} - \dfrac{7}{n^2}$

                                                2._____

3. $\dfrac{y^2}{y+2} + \dfrac{2y}{y+2}$

                                                3._____

4. $\dfrac{m+5}{m^2-49} - \dfrac{12}{m^2-49}$

                                                4._____

5. $\dfrac{c-2b}{c+b} + \dfrac{c+4b}{c+b}$

                                                5._____

6. $\dfrac{w^2+12}{w+1} + \dfrac{9-w^2}{w+1}$

                                                6._____

7. $\dfrac{z^2}{z-2} + \dfrac{z-6}{z-2}$

                                                7._____

8. $\dfrac{11c+71}{10y} - \dfrac{c+1}{10y}$

8._____

9. $\dfrac{y^2}{y^2+2y} - \dfrac{4}{y^2+2y}$

9._____

10. $\dfrac{z^2}{z+4} + \dfrac{4z}{z+4}$

10._____

11. $\dfrac{b-7d}{b+d} + \dfrac{b+9d}{b+d}$

11._____

12. $\dfrac{5z+21}{4c} - \dfrac{z+1}{4c}$

12._____

13. $\dfrac{x^2}{x-14} + \dfrac{x-210}{x-14}$

13._____

14. $\dfrac{h^3-14}{h+7} - \dfrac{2h}{h+7} + \dfrac{7h^2}{h+7}$

14._____

Chapter 7   RATIONAL EXPRESSIONS AND EQUATIONS

7.4 Adding and Subtracting Rational Expressions with Different Denominators

KEY PROPERTIES, PROCEDURES, OR STRATEGIES

Finding the LCD

Adding or Subtracting Rational Expressions with Different Denominators

NOTES

GUIDED EXAMPLE

Subtract $\dfrac{5y}{y^2-13y+36}-\dfrac{4y}{y^2-14y+45}$.

Solution

First, find the LCD by factoring the denominators.

$y^2-13y+36=$ ⎵

$y^2-14y+45=$ ⎵

LCD: ⎵

$\dfrac{5y}{y^2-13y+36}-\dfrac{4y}{y^2-14y+45}$

$=$ ⎵   **Write equivalent rational expressions with the LCD.**

$=$ ⎵   **Distribute in the numerators.**

$=$ ⎵   **Subtract numerators.**

$=$ ⎵   **Factor the numerator.**

$=$ ⎵   **Divide out the common factor.**

166

PRACTICE PROBLEMS

Find the least common denominator for the rational expressions and write equivalent rational expressions with the LCD.

1. $\dfrac{2}{3a^5}, \dfrac{3}{11a^6}$

                                                         1._____

                                                         _____

                                                         _____

2. $\dfrac{2x}{x+4}, \dfrac{7}{x-1}$

                                                          2._____

                                                         _____

                                                         _____

3. $\dfrac{8x}{3x-6}, \dfrac{5x}{4x-8}$

                                                          3._____

                                                         _____

                                                         _____

4. $\dfrac{2}{p^2-4}, \dfrac{7+p}{p+2}$

                                                          4._____

                                                         _____

                                                         _____

Name: Date:
Instructor: Section:

Add or subtract as indicated.

5. $\dfrac{8}{z} + \dfrac{9}{5z}$

5._____

6. $\dfrac{9}{z+3} - \dfrac{5}{z-3}$

6._____

7. $\dfrac{y}{y-c} - \dfrac{c}{c-y}$

7._____

8. $\dfrac{9}{y-5} + \dfrac{7}{5-y}$

8._____

9. $\dfrac{y}{y^2-2y-35} + \dfrac{8}{y+5}$

9._____

10. $\dfrac{x-3}{x+5} + \dfrac{x+6}{x-8}$

10._____

Chapter 7   RATIONAL EXPRESSIONS AND EQUATIONS

7.5 Complex Rational Expressions

KEY VOCABULARY

Term	Definition	Example
Complex rational expression		

KEY PROPERTIES, PROCEDURES, OR STRATEGIES

Simplifying Complex Rational Expressions – Method 1

Simplifying Complex Rational Expressions – Method 2

NOTES

GUIDED EXAMPLE

Simplify $\dfrac{\dfrac{4}{2x+1}}{\dfrac{x}{5x-4}}$.

Solution – Method 1

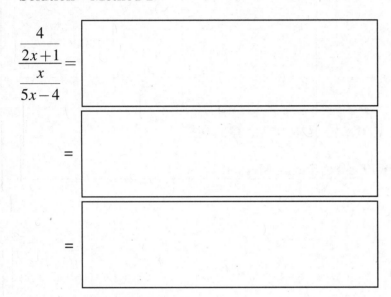

$\dfrac{\dfrac{4}{2x+1}}{\dfrac{x}{5x-4}} =$ ⬚

Rewrite as a horizontal division problem.

$=$ ⬚

Write an equivalent multiplication problem.

$=$ ⬚

Multiply.

Solution – Method 2

$\dfrac{\dfrac{4}{2x+1}}{\dfrac{x}{5x-4}} =$ ⬚

Multiply the numerator and denominator by the LCD:

⬚

$=$ ⬚

Divide out common factors, if necessary.

170

PRACTICE PROBLEMS

Simplify.

1. $\dfrac{-\dfrac{1}{5}}{\dfrac{1}{35}}$

    1._____

2. $\dfrac{\dfrac{t}{u^2}}{\dfrac{t^2}{u}}$

    2._____

3. $\dfrac{\dfrac{8}{7}+\dfrac{3}{14}}{\dfrac{6}{5}-\dfrac{2}{10}}$

    3._____

4. $\dfrac{\dfrac{1}{y}+3}{\dfrac{1}{y}-6}$

    4._____

5. $\dfrac{1+\dfrac{9}{49}}{1-\dfrac{3}{7}}$

    5._____

6. $\dfrac{t - \dfrac{1}{t}}{t + \dfrac{1}{t}}$

6. _____

7. $\dfrac{1 + \dfrac{7}{y}}{1 - \dfrac{49}{y^2}}$

7. _____

8. $\dfrac{\dfrac{9}{x-7} - \dfrac{9}{x+7}}{\dfrac{9}{x^2 - 49}}$

8. _____

9. $\dfrac{\dfrac{1}{y+b} - \dfrac{1}{y}}{b}$

9. _____

10. $\dfrac{1 + \dfrac{2}{y} - \dfrac{3}{y^2}}{\dfrac{1}{y} + \dfrac{3}{y^2}}$

10. _____

Chapter 7   RATIONAL EXPRESSIONS AND EQUATIONS

7.6 Solving Equations Containing Rational Expressions

KEY VOCABULARY

Term	Definition	Example
Extraneous solution		

KEY PROPERTIES, PROCEDURES, OR STRATEGIES

Solving Equations Containing Rational Expressions

NOTES

Name: Date:
Instructor: Section:

GUIDED EXAMPLE
Solve.

$$\frac{1}{3x^2-3}+\frac{2}{x-1}=\frac{3}{x+1}$$

Solution

Inspect the denominators to check for values that make the rational expressions

undefined. Those values are _____ and _____.

Factor the quadratic expression $3x^2-3$:

The LCD of the rational expressions is _____.

$$\left(\right)\left(\frac{1}{3x^2-3}+\frac{2}{x-1}\right)=\left(\frac{3}{x+1}\right)\left(\right)$$ **Multiply both sides by the LCD.**

Distribute and divide out the common factors.

Distribute and combine like terms.

Get the variables on one side of the equation and the constants on the other side.

Solve for x.

Check the solution(s) in the original equation.

Are there any extraneous solutions? _____

Solution(s): _____

174

PRACTICE PROBLEMS

Check the given values to see if they are solutions to the equation.

1. $\dfrac{9}{x} + \dfrac{8}{9} = \dfrac{17}{18}$; 162   1._____

2. $\dfrac{3}{5x} + \dfrac{2}{3} = \dfrac{5}{6x}$; 18   2._____

Solve and check. Identify any extraneous solutions.

3. $\dfrac{a}{4} - \dfrac{a-4}{8} = \dfrac{7}{8}$ 3._____

4. $\dfrac{m}{m-6} - 7 = \dfrac{6}{m-6}$ 4._____

5. $\dfrac{8}{11y} + \dfrac{5}{y} = 1$ 5._____

6. $\dfrac{z+3}{z} = \dfrac{4}{3}$ 6._____

7. $\dfrac{y-6}{y+2}=\dfrac{3}{5}$

7. _____

8. $\dfrac{4}{y+7}=\dfrac{3}{y-6}$

8. _____

9. $\dfrac{3}{z^2-6z+9}-\dfrac{1}{z-3}=\dfrac{1}{6z-18}$

9. _____

10. $\dfrac{1}{2x^2-72}+\dfrac{8}{x-6}=\dfrac{5}{x+6}$

10. _____

11. $\dfrac{6}{6c^2-c-24}=\dfrac{3}{3c^2-3c-8}$

11. _____

12. $\dfrac{6}{y^2-14y+49}-\dfrac{1}{y-7}=\dfrac{1}{8y-56}$

12. _____

Chapter 7   RATIONAL EXPRESSIONS AND EQUATIONS

7.7 Applications with Rational Expressions, Including Variation

KEY VOCABULARY

Term	Definition	Example
Direct variation		
Inverse variation		
Joint variation		

KEY PROPERTIES, PROCEDURES, OR STRATEGIES

Solving Variation Problems

GUIDED EXAMPLE

It takes Holly 2 hours to shovel the snow in her whole driveway by hand. Using a snowplow, Rob can clear the same driveway in $\frac{3}{4}$ hour. If they work together, how long does it take them to clear the driveway of snow?

Solution

Understand   Holly shovels at a rate of 1 driveway in 2 hours, or $\frac{1}{2}$ of a driveway per hour. Rob snowplows at a rate of 1 driveway in $\frac{3}{4}$ hour, or $\frac{1}{\frac{3}{4}} = \frac{4}{3}$ driveways per hour. Let t represent the amount of time it takes Rob and Holly to work together.

Category	Rate of Work	Time at Work	Amount of Task Completed
Holly		t	
Rob		t	

The total job is 1 driveway, so we can write an equation that combines the individual expressions for the tasks completed and set this sum equal to 1 driveway.

Plan and Execute

Holly's amount completed + Rob's amount completed = 1 driveway

Solve the equation for t.

Multiply both sides by the LCD.

Distribute and then divide out the common factors.

Combine like terms.

Solve for t.

Answer   Working together, it takes Rob and Holly _____ to clear the driveway.

Check   Use the value for t to find the amount of the task each person can complete. The sum of these amounts should be 1.

PRACTICE PROBLEMS

Use a table to organize the information; then solve.

1. Carmella can paint a room in 6 hours while it takes Tony 3 hours to paint the same room. How long would it take them to paint the room if they worked together?

1.

2. Cheryl, an experienced shipping clerk, can fill a certain order in 4 hours. Max, a new clerk, needs 5 hours to do the same job. Working together, how long will it take them to fill the order?

2.

3. A steamboat travels 12 kilometers per hour faster   **3.**_____
than a freighter. The steamboat travels 82
kilometers in the same time the freighter travels 50
kilometers. Find the speed of each boat.

4. The speed of a stream is 3 miles per hour. A boat   **4.**_____
travels 7 miles upstream in the same time it takes to
travel 13 miles downstream. What is the speed of
the boat in still water?

Solve the direct variations.

5. If x varies directly as y, and $x = 15$ when $y = 5$, find x when $y = 8$.

5._____

6. The number of kilograms of water in a human body varies directly as the mass of the body. A 90-kilogram person contains 60 kilograms of water. How many kilograms of water are in a 60-kilogram person?

6._____

Solve the inverse variations.

7. The current I in an electrical conductor varies inversely as the resistance R of the conductor. The current is $\dfrac{1}{2}$ ampere when the resistance is 96 ohms. What is the current when the resistance is 216 ohms?

7._____

8. The wavelength W of a radio wave varies inversely as its frequency F. A wave with a frequency of 1400 kilohertz has a length of 150 meters. What is the length of a wave with a frequency of 500 kilohertz?

8._____

Solve the joint variations.

9. Suppose j varies jointly with g and v, and $j = 3$
    when $g = 4$ and $v = 5$. Find j when $g = 10$ and
    $v = 11$.

9._____

10. For a given interest rate, simple interest varies
    jointly as the principal and time. If $2000 left in an
    account for 6 years earned interest of $720, then
    how much interest would be earned in 7 years?

10._____

Solve the combined variation.

11. Suppose that y varies directly with w and the square
    of x and inversely with z, and $y = 4$ when $w = 1$,
    $x = 2$, and $z = 5$. Find y when $w = 4$, $x = 5$, and
    $z = 20$.

11._____

Chapter 8   MORE ON INEQUALITIES, ABSOLUTE VALUE, AND FUNCTIONS

8.1 Compound Inequalities

KEY VOCABULARY

Term	Definition	Example
Compound inequality		
Intersection		
Union		

KEY PROPERTIES, PROCEDURES, OR STRATEGIES

Solving Compound Inequalities Involving *and*

Solving Compound Inequalities Involving *or*

GUIDED EXAMPLES

Solve each inequality and graph the solution set.

a) $10 < 2x + 12 \leq 20$

Solution

$$10 < 2x + 12 \leq 20$$

Subtract 12 from all three parts of the compound inequality.

Divide all three parts of the compound inequality by 2.

Set-builder notation:

Interval notation:

Graph: ⟵――――――――⟶

b) $4x + 5 \geq -19$ or $6x + 2 \geq -10$

Solution

$$4x + 5 \geq -19 \text{ or } 6x + 2 \geq -10$$

$4x + 5 \geq -19$ or $6x + 2 \geq -10$ **Solve each inequality in the compound inequality.**

Isolate the variable term in each inequality.

Solve for x in each inequality.

The solution set is the union of the two individual solution sets.

Set-builder notation:

Interval notation:

Graph: ⟵――――――――⟶

184

PRACTICE PROBLEMS

a. *Write the solution set using set-builder notation.*
b. *Write the solution set using interval notation.*
c. *Graph the solution set.*

1. $x < 2$ and $x > -2$

1._____

_____

⟵——————————⟶

2. $x + 8 \geq 6$ and $x + 7 \leq 9$

2._____

_____

⟵——————————⟶

3. $x \geq 6$ or $x \geq 8$

3._____

_____

⟵——————————⟶

4. $x > -4$ or $x \leq 4$

4._____

_____

⟵——————————⟶

5. $-3 < x + 1 < 2$

5._____

_____

⟵——————————⟶

6. $11 < 3x + 14 \leq 23$

6._____

_____

⟵——————————⟶

185

7. $-2x-2<-10$ or $-2x-2>10$

7._____

_____

<—————————————>

8. $9x+9\geq-45$ or $3x+1\geq-5$

8._____

_____

<—————————————>

9. $0\leq x+4$ and $x+4<10$

9._____

_____

<—————————————>

10. $5x-1>-16$ or $5x-1\leq5$

10._____

_____

<—————————————>

11. $-4<2-2x\leq10$

11._____

_____

<—————————————>

12. $8x+1\geq-71$ or $5x+3\geq-12$

12._____

_____

<—————————————>

Chapter 8   MORE ON INEQUALITIES, ABSOLUTE VALUE, AND FUNCTIONS

8.2 Equations Involving Absolute Value

KEY PROPERTIES, PROCEDURES, OR STRATEGIES

Absolute Value Property

In the Language of Math	In Your Own Words

Solving Equations Containing a Single Absolute Value

Solving Equations in the Form $|ax + b| = |cx + d|$

NOTES

GUIDED EXAMPLES

1. Solve: $|6x - 3| = 15$

Solution

$|6x - 3| = 15$

$6x - 3 = 15$ or $6x - 3 = -15$ Separate the equation into a positive case and a negative case.

Solve each case separately. Isolate the variable term.

Solve for x.

The solutions are ☐ and ☐.

2. Solve: $|9x + 4| = |x - 13|$

Solution

$|9x + 4| = |x - 13|$

Equal   **Opposites**   Separate the absolute value equation into two equations.

$9x + 4 = x - 13$ or $9x + 4 = -(x - 13)$

Solve each case separately. Distribute the negative sign in the second equation.

Isolate the variable term.

Solve for x.

The solutions are ☐ and ☐.

188

PRACTICE PROBLEMS

Solve using the absolute value property.

1. $|x| = 9$

    1._____

2. $|x| = -19$

    2._____

3. $|x - 6| = 14$

    3._____

4. $|4x - 3| = 6$

    4._____

5. $|4x - 4| = 0$

    5._____

6. $\left|\frac{1}{2}x + 2\right| = 4$

    6._____

Isolate the absolute value; then use the absolute value property.

7. $|6x| - 3 = 33$

    7._____

189

8. $|x+8|+2=31$

8._____

9. $|7x+3|-8=-2$

9._____

10. $12-|x+7|=4$

10._____

11. $|x+4|+3=18$

11._____

Solve by separating the two absolute values into equal and opposite cases.

12. $|3x+8|=|x-11|$

12._____

13. $|x-15|=|15-x|$

13._____

14. $|3x+8|=|x-17|$

14._____

Chapter 8   MORE ON INEQUALITIES, ABSOLUTE VALUE, AND FUNCTIONS

8.3 Inequalities Involving Absolute Value

KEY PROPERTIES, PROCEDURES, OR STRATEGIES

Solving Inequalities in the Form $|x| < a$, where $a > 0$

Solving Inequalities in the Form $|x| > a$, where $a > 0$

NOTES

GUIDED EXAMPLES

For each inequality, solve, graph the solution set, and write the solution set in both set-builder and interval notation.

a) $|x+5|+8 < 28$

Solution

$$|x+5|+8 < 28$$

Subtract 8 from both sides to isolate the absolute value.

Rewrite as a compound inequality.

Subtract 5 from all three parts of the inequality.

Graph:

Set-builder notation:

Interval notation:

b) $|x+19| > 9$

Solution

$$|x+19| > 9$$

Rewrite as a compound inequality.

Subtract 19 from both sides of each inequality.

Graph:

Set-builder notation:

Interval notation:

192

PRACTICE PROBLEMS

Solve the inequality. Then:
  a.   *Write the solution set using set-builder notation.*
  b.   *Write the solution set using interval notation.*
  c.   *Graph the solution set.*

1. $|x| < 5$

1._____

_____

$\longleftrightarrow$

2. $|x + 5| \le 4$

2._____

_____

$\longleftrightarrow$

3. $|x + 2| + 8 \le 11$

3._____

_____

$\longleftrightarrow$

4. $2|x| - 2 \le 4$

4._____

_____

$\longleftrightarrow$

5. $|x| > 2$

5._____

_____

$\longleftrightarrow$

6. $|x+7|>6$

6._____

_____

←——————————→

7. $5-|x+3|>9$

7._____

_____

←——————————→

8. $3|x|+4\geq10$

8._____

_____

←——————————→

9. $|5x-5|>-23$

9._____

_____

←——————————→

10. $9-|x+3|>-1$

10._____

_____

←——————————→

Chapter 8   MORE ON INEQUALITIES, ABSOLUTE VALUE, AND FUNCTIONS

8.4 Functions and Graphing

KEY VOCABULARY

Term	Definition	Example
Relation		
Domain		
Range		
Function		

NOTES

GUIDED EXAMPLES
Graph.

a) $f(x) = 2x + 5$

   Solution

Use the *y*-intercept as one ordered pair on the graph

of the line:

Use the slope of the line to find a second ordered

pair on the graph of the line:

b) $f(x) = -x^2 - 1$

   Solution

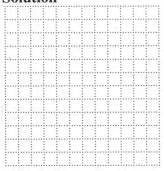

Recall that the sign of *a* in $f(x) = ax^2 + bx + c$ indicates whether the parabola opens up or down. Create a table of ordered pairs, plot the points, and connect them with a smooth curve.

x	*f(x)*

c) $f(x) = |x + 2| - 3$

   Solution

Create a table of ordered pairs, plot the points, and connect them. Consider how the graph of $f(x) = |x|$ is shifted to create this graph.

x	*f(x)*

Name: _____   Date: _____

Instructor: _____   Section: _____

PRACTICE PROBLEMS

Identify the domain and range of each relation and determine whether it is a function.

1. $\{(12,2),(27,-10),(34,2),(12,6),(48,6)\}$

1._____

_____

_____

2.

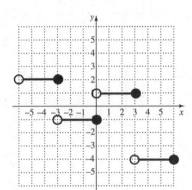

2._____

_____

_____

3.

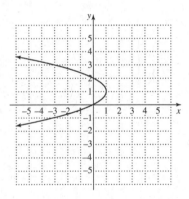

3._____

_____

_____

Find each of the following.

4. Given $f(x) = 10x^2 + 9x - 8$, find $f(6)$.

4._____

5. Given $f(x) = \dfrac{x-12}{7x-11}$, find $f(-5)$.

5._____

6. Given $f(x) = -2x^3 + 4x^2 - 1$, find $f(2)$.

6._____

197

Use the graph to determine the value of the function.

7. $f(2)$ **7.**_____

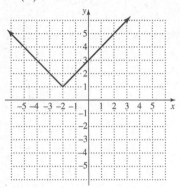

Graph.

8. $f(x) = -4x - 3$ **8.**

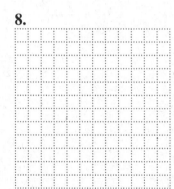

9. $f(x) = -x^2 + 3$ **9.**

10. $f(x) = |x - 3| + 4$ **10.**

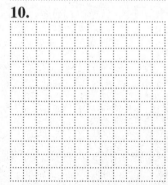

Chapter 8   MORE ON INEQUALITIES, ABSOLUTE VALUE, AND FUNCTIONS

8.5 Function Operations

KEY PROPERTIES, PROCEDURES, OR STRATEGIES

Adding or Subtracting Functions

In the Language of Math	In Your Own Words

Multiplying Functions

In the Language of Math	In Your Own Words

Dividing Functions

In the Language of Math	In Your Own Words

GUIDED EXAMPLES

Given $f(x) = x^2 - 4$ and $g(x) = x + 2$, find the following.

a) $f + g$

    Solution

$$(f+g)(x) = f(x) + g(x)$$

$$= \boxed{} \quad \text{Write a sum of the two functions.}$$

$$= \boxed{} \quad \text{Combine like terms.}$$

b) $f - g$

Solution

$(f - g)(x) \;=\; f(x) - g(x)$

$=\;$ [ ]   Write a difference of the two functions.

$=\;$ [ ]   Write as an equivalent addition.

$=\;$ [ ]   Combine like terms.

c) $f \cdot g$

Solution

$(f \cdot g)(x) \;=\; f(x)g(x)$

$=\;$ [ ]   Write a product of the two functions.

$=\;$ [ ]   Use FOIL to multiply the binomials.

$=\;$ [ ]   Combine like terms.

d) f / g

Solution

$(f / g)(x) \;=\; \dfrac{f(x)}{g(x)}$

$=\;$ [ ]   Write a quotient of the two functions.

$=\;$ [ ]   Factor the numerator.

$=\;$ [ ]   Divide.

200

PRACTICE PROBLEMS

Find $f+g$ and $f-g$.

1. $f(x)=4x^2$, $g(x)=8x-7$

1._____

_____

2. $f(x)=-5x+3$, $g(x)=x^2+2$

2._____

_____

3. $f(x)=x+2$, $g(x)=x-2$

3._____

_____

4. $f(x)=x-1$, $g(x)=x^2-x$

4._____

_____

5. $f(x)=2x^2+3$; $g(x)=x^3+6x$

5._____

_____

6. $f(x)=-4x^2+3x+2$, $g(x)=7x^2-x+5$

6._____

_____

7. $f(x)=2x^2+6x-1$, $g(x)=2x^2+3x-4$

7._____

_____

8. $f(x)=x^3-2x+4$, $g(x)=2x^3-x^2+7x$

8._____

_____

Find $f \cdot g$.

9. $f(x) = 2x - 9$, $g(x) = 3x + 1$ 9._____

10. $f(x) = x^2 - 18$, $g(x) = 14 - x$ 10._____

11. $f(x) = x^2 - 3x + 2$, $g(x) = x^2 + 2x - 9$ 11._____

Find f / g.

12. $f(x) = x^2 - 7x + 10$, $g(x) = x - 2$ 12._____

13. $f(x) = x^2 - 1$, $g(x) = x + 1$ 13._____

14. $f(x) = 5x^3 - 10x^2$, $g(x) = 5x$ 14._____

Chapter 9   SYSTEMS OF LINEAR EQUATIONS AND INEQUALITIES

9.1 Solving Systems of Linear Equations Graphically

KEY VOCABULARY

Term	Definition	Example
System of equations		
Solution for a system of equations		
Consistent system of equations		
Inconsistent system of equations		

KEY PROPERTIES, PROCEDURES, OR STRATEGIES

Checking a Solution to a System of Equations

Solving Systems of Linear Equations Graphically

Classifying Systems of Equations

GUIDED EXAMPLE

Determine whether the system is consistent with independent equations, consistent with dependent equations, or inconsistent. Discuss the number of solutions for the system.

$$\begin{cases} x - y = 2 \\ 9x - 3y = 45 \end{cases}$$

Solution

Write the equations in slope-intercept form.

$x - y = 2$ | $9x - 3y = 45$

Are the slopes the same or different? _____

Do the lines have the same *y*-intercept? _____

Classify the system as consistent with independent equations, consistent with dependent

equations, or inconsistent. _____

How many solutions does the system have? _____

PRACTICE PROBLEMS

Determine whether the given ordered pair is a solution to the given system of equations.

1. $(-4,-2)$; $\begin{cases} 3x+5y=-22 \\ 4x+2y=-20 \end{cases}$ 1._____

2. $(7,4)$; $\begin{cases} 3x-2y=13 \\ 4x-9y=11 \end{cases}$ 2._____

Solve the system of linear equations graphically.

3. $\begin{cases} x+y=10 \\ x-y=-2 \end{cases}$ 3._____

4. $\begin{cases} y=2x-2 \\ 3x+y=3 \end{cases}$ 4._____

5. $\begin{cases} 3x-y=3 \\ 2x+y=2 \end{cases}$ 5._____

6. $\begin{cases} 2x - 6y = 6 \\ 3x - 9y = -6 \end{cases}$ 6._____

7. $\begin{cases} 4x - 5y = 2 \\ 10y = 8x - 4 \end{cases}$ 7._____

a. *Determine whether the system of equations is consistent with independent equations, consistent with dependent equations, or inconsistent.*

b. *Determine how many solutions the system has.*

8. $\begin{cases} x - y = 6 \\ 6x - 6y = 37 \end{cases}$ 8a._____

                                                          b._____

9. $\begin{cases} 7x + y = 4 \\ 35x + 5y = 20 \end{cases}$ 9a._____

                                                          b._____

10. $\begin{cases} 3x + y = 13 \\ 4x + 5y = -1 \end{cases}$ 10a._____

                                                          b._____

Name: Date:
Instructor: Section:

Chapter 9   SYSTEMS OF LINEAR EQUATIONS AND INEQUALITIES

9.2 Solving Systems of Linear Equations by Substitution; Applications

KEY PROPERTIES, PROCEDURES, OR STRATEGIES

Solving Systems of Two Linear Equations Using Substitution

Solving Applications Using a System of Equations

NOTES

GUIDED EXAMPLE

Solve the system of equations using substitution.

$$\begin{cases} x - y = 3 \\ 7x + 3y = -39 \end{cases}$$

Solution

Step 1: Isolate a variable in one of the equations. Because the variables have coefficients of 1 and -1 in $x - y = 3$, isolating either variable is easy. We will isolate x.

$$x - y = 3$$

$$x = \boxed{}$$

Step 2: Substitute the expression for x found in step 1 for x in the second equation.

$$7x + 3y = -39$$

$$7\left(\right) + 3y = -39$$

Step 3: Solve for y using the equation found in step 2.

$\boxed{}$	**Distribute 7.**
$\boxed{}$	**Combine like terms.**
$\boxed{}$	**Isolate the variable term.**
$\boxed{}$	**Solve for y.**

Step 4: Solve for x by substituting the value found for y in one of the equations containing both variables.

Solution: $\boxed{}$

Step 5: Check the solution in the original equations.

GUIDED EXAMPLE

There were 334 tickets sold for a basketball game. Student tickets cost $1.50 and non-student tickets cost $2.00. The total amount of money collected was $527.00. How many of each kind of ticket were sold?

Solution

Understand The two unknowns are the number of student tickets sold and the number of non-student tickets sold. One relationship involves the number of tickets sold (334 total), and the other relationship involves the total sales in dollars ($527).

Plan and Execute

Let x represent the number of student tickets sold and y represent the number of non-student tickets sold.

Relationship 1:   The total number of tickets sold was 334. Write an equation to represent this relationship.

$$\boxed{}$$

Relationship 2:   The total sales were $527. Write an equation to represent this relationship.

$$\boxed{}$$

Solve the system using the substitution method.

Answer Number of student tickets sold: _____

Number of non-student tickets sold: _____

Check   Verify the solution in both given relationships.

GUIDED EXAMPLE

A freight train leaves a station and travels north at 60 miles per hour. Four hours later, a passenger train leaves on a parallel track and travels north at 100 miles per hour. How long will it take the passenger train to overtake the freight train? How far from the station will they be at this time?

Solution

Understand To determine the time it takes for the passenger train to overtake the freight train, use a table to organize the rates and times.

Plan and Execute   Let t represent the time for the passenger train to catch up to the freight train. Add 4 hours to t to represent the freight train's time. Use the fact that distance = rate × time.

Trains	Rate	Time	Distance d
Freight train			$d =$
Passenger train			$d =$

Write and solve the system of equations. Use the substitution method.

Length of time the freight train has traveled: _____ hours

Distance the freight train has traveled: _____ miles

Length of time the passenger train has traveled: _____ hours

Distance the passenger train has traveled: _____ miles

Answer   It will take the passenger train _____ hours to catch up to the freight train. At that time, the trains have each traveled _____ miles.

Check   Verify that the trains have traveled the same distance after each train has traveled its respective length of time.

PRACTICE PROBLEMS

Solve the system of equations using substitution. Note that some systems may be inconsistent or consistent with dependent equations.

1. $\begin{cases} x + y = 4 \\ y = x - 2 \end{cases}$

1._____

2. $\begin{cases} 9x + 2y = -55 \\ x = 17 - 6y \end{cases}$

2._____

3. $\begin{cases} x - 9y = 62 \\ 4y - 7x = -80 \end{cases}$

3._____

4. $\begin{cases} 3x + y = 24 \\ y = 3x \end{cases}$

4._____

5. $\begin{cases} x - 2 = y \\ 4x - 4y = 8 \end{cases}$

5._____

6. $\begin{cases} 2x - 5y = 5 \\ 2x + 5y = 5 \end{cases}$

6._____

7. $\begin{cases} 4x + 5y = 14 \\ 2x + 3y = 10 \end{cases}$

7._____

8. $\begin{cases} x - 9 = y \\ 9x - 9y = 45 \end{cases}$

8._____

Translate the problem to a system of equations and then solve.

9. The sum of two numbers is 35. One number is 9 more than the other. Find the numbers.

9._____

10. Two angles are supplementary. One angle is $3°$ smaller than twice the other. Find the measures of the angles.

10._____

Solve.

11. A motel clerk counts his $1 and $10 bills at the end of a day. He finds that he has a total of 45 bills having a combined monetary value of $162. Find the number of bills of each denomination that he has.

11._____

12. The Everton College store paid $1665 for an order
of 45 calculators. The store paid $9 for each
scientific calculator. The others, all graphing
calculators, cost the store $54 each. How many of
each type of calculator was ordered?

12._____

13. A train leaves a station and travels north at 75
kilometers per hour. Two hours later, a second train
leaves on a parallel track and travels north at 115
kilometers per hour. How far from the station will
they meet?

13._____

Chapter 9   SYSTEMS OF LINEAR EQUATIONS AND INEQUALITIES

9.3 Solving Systems of Linear Equations by Elimination; Applications

KEY PROPERTIES, PROCEDURES, OR STRATEGIES

Solving Systems of Two Linear Equations Using Elimination

NOTES

GUIDED EXAMPLE

Solve the system of equations using the elimination method.

$$\begin{cases} 4x - 2y = 38 \\ 5x + 3y = -2 \end{cases}$$

Solution

The equations are already in standard form and there are no fractions or decimals to be cleared.

We must multiply both equations by numbers that create a pair of terms that are additive inverses. This can be done in an infinite number of ways, but here we will multiply the first equation by 3 and the second equation by 2 so that the y terms are additive inverses.

$$4x - 2y = 38 \qquad \xrightarrow{\text{Multiply by 3.}} \qquad \boxed{}$$

$$5x + 3y = -2 \qquad \xrightarrow{\text{Multiply by 2.}} \qquad \boxed{}$$

Now we can add the rewritten equations to eliminate the y term.

$\boxed{}$ **Equation 1 rewritten**

$+\ \underline{\boxed{}}$ **Equation 2 rewritten**

$\boxed{}$ **Add the equations to eliminate y.**

$\boxed{}$ **Isolate x.**

To finish, substitute the value for x in one of the equations and solve for y.

Solution: $\boxed{}$

Check the solution in the original equations.

Name: Date:

Instructor: Section:

GUIDED EXAMPLE

A pharmacist has a 40% acid solution and a 25% acid solution. How many liters of each must be mixed to form 105 liters of a 27% acid solution?

Solution

Understand There are three solutions in this problem: the 40% solution, the 25% solution, and the new 27% solution that is created by mixing the 40% and 25% solutions.

Plan and Execute   We are to find the volume of the 40% solution and the volume of the 25% solution. Choose variables to represent each of these unknown amounts, and use the fact that we want to form 105 liters of a 27% acid solution

Solutions	Concentration of Acid	Volume of Solution	Volume of Acid
40% solution			
25% solution			
27% solution			

Write an equation that describes the total volume: _____

Write an equation that describes the relation ship between the volumes of acid:

_____

Use the elimination method to solve this system of equations.

Answer   Volume of 40% acid solution needed: _____ liters

Volume of 25% acid solution needed: _____ liters

Check   Verify that the volume of acid in the two original solutions combined is equal to the volume of acid in the combined solution.

Name: Date:
Instructor: Section:

NOTES

PRACTICE PROBLEMS

Solve the system of equations using the elimination method. Note that some systems may be inconsistent or consistent with dependent equations.

1. $\begin{cases} x + y = 3 \\ x - y = 5 \end{cases}$ 1._____

2. $\begin{cases} x + y = 15 \\ 8x - y = 57 \end{cases}$ 2._____

3. $\begin{cases} 2x + 5y = 14 \\ 9x - 5y = -102 \end{cases}$ 3._____

4. $\begin{cases} 3x - 4y = 26 \\ 2x + 5y = 2 \end{cases}$ 4._____

5. $\begin{cases} 5x + y = -18 \\ 7x - 3y = -56 \end{cases}$ 5._____

6. $\begin{cases} 3x+4y=1 \\ 6x+8y=2 \end{cases}$

6._____

7. $\begin{cases} \dfrac{5}{2}x+\dfrac{5}{3}y=15 \\ \dfrac{1}{4}x+\dfrac{1}{3}y=2 \end{cases}$

7._____

8. $\begin{cases} 0.5x=-45.5+3y \\ -\dfrac{6}{5}x+y=\dfrac{19}{5} \end{cases}$

8._____

Translate the problem to a system of equations. Then solve using the elimination method.

9. Two angles are supplementary. One is $90°$ more than twice the other. Find the measures of the angles.

9._____

10. The perimeter of a rectangular rug is 28 feet. The length is 2 feet longer than the width. Find the dimensions of the rug.

10._____

Solve.

11. If a plane can travel 490 miles per hour with the wind and 410 miles per hour against the wind, find the speed of the plane in still air.

11._____

12. Mr. White invested $29,000 in two accounts, one yielding 6% interest and the other yielding 10%. If he received a total of $2100 in interest at the end of the year, how much did he invest in each account?

12._____

13. Sheila Grant has twice as much money invested at 5% simple annual interest as she does at 4%. If her yearly income from the two investments is $119, how much does she have at each rate?

13._____

14. One canned juice drink is 25% orange juice; another is 10% orange juice. How many liters of each should be mixed together in order to get 15 liters that is 13% orange juice?

14._____

15. A store mixes Kenyan coffee worth $12 per
kilogram and Turkish coffee worth $15 per
kilogram. The mixture is to sell for $13 per
kilogram. Find how much of each should be used to
make a 588-kilogram mixture.

15. _____

16. A mother wants to invest $15,000 for her son's
future education. She invests a portion of the money
in a bank certificate of deposit (CD account) which
earns 4% and the remainder in a savings bond that
earns 7%. If the total interest earned after one year
is $900, how much money was invested in the CD
account?

16. _____

17. A chocolate factory makes dark chocolate that is
33% fat and a white chocolate that is 47% fat. How
many kilograms of dark chocolate should be mixed
with 700 kilograms of white chocolate to make a
ripple blend that is 35% fat?

17. _____

Chapter 9 SYSTEMS OF LINEAR EQUATIONS AND INEQUALITIES

9.4 Solving Systems of Linear Equations in Three Variables; Applications

KEY PROPERTIES, PROCEDURES, OR STRATEGIES

Solving Systems of Three Linear Equations Using Elimination

NOTES

GUIDED EXAMPLE

Solve the system of equations using elimination.

$$\begin{cases} 3x+3y+z=21 \\ x-3y+2z=-5 \\ 8x-2y+3z=26 \end{cases}$$

Solution

Choose a variable and eliminate that variable from two pairs of equations. Let's eliminate x using equations 1 and 2.

(Eq. 1) $3x+3y+z=21$ $3x+3y+z=21$

(Eq. 2) $x-3y+2z=-5$ $\xrightarrow{\text{Multiply by }-3.}$ ⬚   **Add the equations.**

⬚   (Equation 4)

Now eliminate x from equations 2 and 3.

(Eq. 2) $x-3y+2z=-5$ $\xrightarrow{\text{Multiply by }-8.}$ ⬚

(Eq. 3) $8x-2y+3z=26$ $8x\ -\ 2y\ +\ 3z\ =\ 26$ **Add the equations.**

⬚   (Equation 5)

Equations 4 and 5 form a system of two equations in two variables y and z. Solve these equations using the elimination method.

⬚

To solve for x, substitute the values of y and z into any of the three original equations.

⬚

Check the ordered triple in all three original equations.

PRACTICE PROBLEMS

Determine whether the ordered triple is a solution of the system.

1. $(-6, 3, -3)$

$$\begin{cases} x + y + z = -6 \\ x - 2y - z = -9 \\ 2x + 2y - z = -3 \end{cases}$$

1._____

2. $(5, 1, -2)$

$$\begin{cases} x + y + z = 4 \\ x - 2y - z = 5 \\ 3x + 2y - z = 15 \end{cases}$$

2._____

Solve the system of equations.

3.
$$\begin{cases} x + y + z = -2 \\ 2x + 5y + 2z = -10 \\ -x + 6y - 3z = -16 \end{cases}$$

3._____

4.
$$\begin{cases} 4x + 3y + z = 4 \\ x - 3y + 2z = 9 \\ 11x - 2y + 3z = 51 \end{cases}$$

4._____

5.
$$\begin{cases} 7x - y + z = -25 \\ 2x + 2y - 3z = 6 \\ x - 3y + 2z = -11 \end{cases}$$

5._____

6.
$$\begin{cases} 5x + 5y - 10z = 7 \\ 3x + y - z = 5 \\ -x - y + 2z = 3 \end{cases}$$

6._____

7.
$$\begin{cases} -3x + 3y - 9z = -12 \\ x - y + 3z = 4 \\ 5x - 5y + 15z = 20 \end{cases}$$

7._____

8.
$$\begin{cases} x = -y - z - 5 \\ 2x + 5y + 2z = 2 \\ -x + 7y - 3z = 47 \end{cases}$$

8._____

9.
$$\begin{cases} 2x - 3y + 4z = 4 \\ x + 2y + z = 14 \\ 5y - z = 18 \end{cases}$$

9._____

10.
$$\begin{cases} \dfrac{7}{2}x + y - z = 0 \\ 2y - 4z = -72 \\ -0.4x + 0.2y = -3.6 \end{cases}$$

10._____

Name: Date:

Instructor: Section:

Translate to a system of three equations; then solve.

11. The sum of the measures of the angles in every triangle is $180°$. In triangle *ABC*, the measure of angle *B* is $26°$ more than three times the measure of angle *A*. The measure of angle *C* is $49°$ more than the measure of angle *A*. Find the measure of each angle.

11._____

12. A basketball team recently scored a total of 80 points on a combination of 2-point field goals, 3-point field goals, and 1-point foul shots. Altogether, the team made 45 baskets and 19 more 2-point field goals than foul shots. How many shots of each kind were made?

12._____

Chapter 9 SYSTEMS OF LINEAR EQUATIONS AND INEQUALITIES

9.5 Solving Systems of Linear Equations Using Matrices

KEY VOCABULARY

Term	Definition	Example
Matrix		
Augmented matrix		
Row echelon form		

KEY PROPERTIES, PROCEDURES, OR STRATEGIES

Row Operations

NOTES

GUIDED EXAMPLE

Solve the following linear system by transforming the augmented matrix into row echelon form.

$$\begin{cases} -15x - 3y = -54 \\ 6x + y = 22 \end{cases}$$

Solution

First, write the augmented matrix.

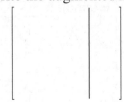

Now perform row operations to get row echelon form.

$$-\frac{1}{15}R_1 \rightarrow \begin{bmatrix} & & \\ & & \end{bmatrix}$$

We need a 1 in row 1, column 1, so multiply row 1 by $-\dfrac{1}{15}$.

$$\begin{bmatrix} & & \\ & & \end{bmatrix}$$

$$-6R_1 + R_2 \rightarrow \begin{bmatrix} & & \\ & & \end{bmatrix}$$

To get a 0 in row 2, column 1, we multiply the new row 1 by –6 and add it to row 2.

$$-5R_2 \rightarrow \begin{bmatrix} & & \\ & & \end{bmatrix}$$

To get a 1 in row 2, column 2, we multiply row 2 by –5.

Write the system that the resulting matrix represents.

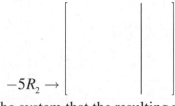

We can solve for x using substitution.

The solution is ☐ .

PRACTICE PROBLEMS

Write the augmented matrix for the system of equations.

1. $\begin{cases} -x+6y=5 \\ 3x-5y=7 \end{cases}$ 1._____

2. $\begin{cases} 2x-8y+3z=-5 \\ 5x-4y+2z=9 \\ 4y-6z=-9 \end{cases}$ 2._____

Given the matrices in row echelon form, find the solution for the system.

3. $\begin{bmatrix} 1 & 4 & | & -3 \\ 0 & 1 & | & 7 \end{bmatrix}$ 3._____

4. $\begin{bmatrix} 1 & 1 & -1 & | & 11 \\ 0 & 1 & -1 & | & 2 \\ 0 & 0 & 1 & | & 1 \end{bmatrix}$ 4._____

Complete the indicated row operation.

5. Replace R_1 in $\begin{bmatrix} 16 & 12 & | & 4 \\ 0 & 3 & | & 9 \end{bmatrix}$ with $\dfrac{1}{4}R_1$.   5._____

6. Replace R_2 in $\begin{bmatrix} 1 & 7 & 4 & | & 0 \\ -6 & 8 & -1 & | & 4 \\ 5 & 7 & 0 & | & 2 \end{bmatrix}$ with $6R_1+R_2$.   6._____

Solve by transforming the augmented matrix into row echelon form.

7. $\begin{cases} x+y=2 \\ x-y=6 \end{cases}$ 7._____

8. $\begin{cases} -25x - 3y = -116 \\ 10x + y = 47 \end{cases}$ 8._____

9. $\begin{cases} x + y - z = 6 \\ 2x - y + z = -12 \\ x - 4y + 3z = -46 \end{cases}$ 9._____

10. $\begin{cases} -35x - 2y + z = -97 \\ 20x + y = 57 \\ -40x - 2y + z = -112 \end{cases}$ 10._____

Solve using a matrix on a graphing calculator.

11. $\begin{cases} 3x + y = 19 \\ 5x + y = 29 \end{cases}$ 11._____

12. $\begin{cases} -21x - 2y + z = -73 \\ 12x + y = 44 \\ -24x - 2y + z = -85 \end{cases}$ 12._____

13. $\begin{cases} 5x + y - 4z = -12 \\ 5x + 5y + z = -20 \\ -3x - 5y + 5z = 16 \end{cases}$ 13._____

Translate the problem to a system of equations; then solve using matrices.

14. Two rivers have a combined length of 5920 miles.   14._____
The longer of the two rivers is 220 miles longer
than the shorter. What is the length of each river?

Chapter 9 SYSTEMS OF LINEAR EQUATIONS AND INEQUALITIES

9.6 Solving Systems of Linear Equations Using Cramer's Rule

KEY VOCABULARY

Term	Definition	Example
Square matrix		
Minor of an element of a matrix		

KEY PROPERTIES, PROCEDURES, OR STRATEGIES

Determinant of a 2 × 2 matrix

In the Language of Math	In Your Own Words

Evaluating the Determinant of a 3 × 3 Matrix

In the Language of Math	In Your Own Words

Cramer's Rule

GUIDED EXAMPLE

Find the determinant of $\begin{bmatrix} 6 & 2 & 10 \\ -3 & 1 & 10 \\ 8 & 7 & 0 \end{bmatrix}$.

Solution

Using the rule for expanding by minors along the first column, we have:

$$\begin{vmatrix} 6 & 2 & 10 \\ -3 & 1 & 10 \\ 8 & 7 & 0 \end{vmatrix} = 6\begin{pmatrix} \text{minor} \\ \text{of } 6 \end{pmatrix} - (-3)\begin{pmatrix} \text{minor} \\ \text{of } -3 \end{pmatrix} + 8\begin{pmatrix} \text{minor} \\ \text{of } 8 \end{pmatrix}$$

$$= 6\begin{vmatrix} & \\ & \end{vmatrix} - (-3)\begin{vmatrix} & \\ & \end{vmatrix} + 8\begin{vmatrix} & \\ & \end{vmatrix}$$

$$=$$

$$=$$

$$=$$

234

Name: Date:
Instructor: Section:

PRACTICE PROBLEMS

Find the determinant.

1. $\begin{bmatrix} 1 & 9 \\ -6 & -4 \end{bmatrix}$

1._____

2. $\begin{bmatrix} -3 & -8 \\ -9 & 3 \end{bmatrix}$

2._____

3. $\begin{bmatrix} 9 & 6 & -2 \\ 6 & 5 & 1 \\ 2 & -3 & 5 \end{bmatrix}$

3._____

4. $\begin{bmatrix} -1 & 8 & 2 \\ 0 & 5 & 5 \\ 0 & 2 & 3 \end{bmatrix}$

4._____

5. $\begin{bmatrix} \dfrac{1}{3} & -\dfrac{4}{5} & \dfrac{3}{5} \\ \dfrac{1}{4} & \dfrac{1}{5} & -\dfrac{4}{3} \\ -\dfrac{4}{5} & \dfrac{1}{3} & \dfrac{4}{5} \end{bmatrix}$

5._____

6. $\begin{bmatrix} x & y & 1 \\ 4 & -5 & 1 \\ -3 & 0 & 2 \end{bmatrix}$

6._____

235

Name: Date:
Instructor: Section:

Solve using Cramer's rule.

7. $\begin{cases} 6x + 2y = 1 \\ 5x + 8y = 10 \end{cases}$

7._____

8. $\begin{cases} 8x + 7y = 16 \\ 8x + 6y = 16 \end{cases}$

8._____

9. $\begin{cases} \dfrac{2}{5}x + \dfrac{4}{5}y = -\dfrac{5}{3} \\[2mm] \dfrac{3}{5}x + \dfrac{3}{5}y = -\dfrac{2}{7} \end{cases}$

9._____

10. $\begin{cases} 4x + 4y + z = 8 \\ 3x - y + z = 6 \\ -x + y - 3z = -2 \end{cases}$

10._____

11. $\begin{cases} 4x + 3z = 0 \\ 5x + 2y = 21 \\ -y + 5z = -48 \end{cases}$

11._____

12. $\begin{cases} 0.4x + 0.3y - 0.2z = 1.9 \\ 0.1x - 0.3y + 0.6z = 0 \\ 2.9x + 1.9y - 2.2z = 12.2 \end{cases}$

12._____

Chapter 9   SYSTEMS OF LINEAR EQUATIONS AND INEQUALITIES

9.7 Solving Systems of Linear Inequalities

KEY PROPERTIES, PROCEDURES, OR STRATEGIES

Solving a System of Linear Inequalities in Two Variables

GUIDED EXAMPLE
Graph the solution set for the system of inequalities.

$$\begin{cases} 3x + y \geq 2 \\ x \leq 3 \end{cases}$$

Solution
Graph the inequalities on the same grid. Both lines should be [solid / dashed].

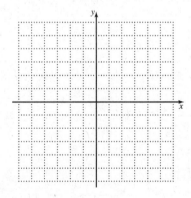

The solution set for this system contains (circle all that apply):

All ordered pairs in the region of overlap

All ordered pairs on the line $3x + y = 2$ that touch the region of overlap

All ordered pairs on the line $x = 3$ that touch the region of overlap

237

Name: _____   Date: _____
Instructor: _____   Section: _____

GUIDED EXAMPLE

A company sells a regular version of software for $15.95 and a deluxe version for $20.95. The company gives a bonus to any salesperson who sells at least 100 units and has at least $1900 in total sales. Write a system of inequalities that describes the requirements a salesperson must meet in order to receive the bonus. Solve the system by graphing.

Solution

Understand We must translate to a system of inequalities, then solve the system.

Plan and Execute   Let x represent the number of regular units sold and y represent the number of deluxe units.

Relationship 1:   The salesperson must sell at least 100 units, so the total number of units sold must be greater than or equal to 100. The first inequality in the system is

Relationship 2:   The salesperson must have at least $1900 in total sales, earned by selling x units at $15.95 and y units at $20.95. So the second inequality in the system is

Answer   Now graph each inequality separately. Because the salesperson cannot sell a negative number of units, the solution set is confined to quadrant I. Any ordered pair in the solution region or on a portion of either line touching the solution region is a solution for the system. However, assuming that only whole units can be sold, only ordered pairs of whole numbers in the solution set are realistic.

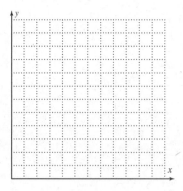

Check   Check an ordered pair in the solution region.

238

Name:

Instructor:

Date:

Section:

PRACTICE PROBLEMS

Graph the solution set for the system of inequalities.

1. $\begin{cases} 5x + 3y < 15 \\ x - y < 5 \end{cases}$

1.

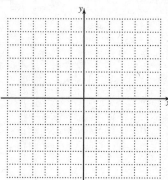

2. $\begin{cases} 2x + 5y \geq 10 \\ 3x - 2y \leq 6 \end{cases}$

2.

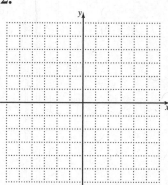

3. $\begin{cases} y > x \\ y < -x + 3 \end{cases}$

3.

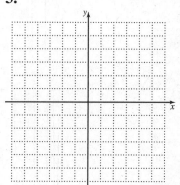

239

4. $\begin{cases} 2x > -3y \\ 2x + 3y \le -6 \end{cases}$

4.

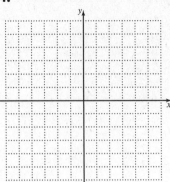

5. $\begin{cases} x + y \le 2 \\ x - y \le 4 \end{cases}$

5.

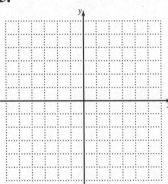

6. $\begin{cases} y < 3x + 1 \\ 3x - y \le 5 \end{cases}$

6.

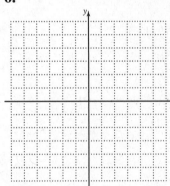

7. $\begin{cases} 4x + 4y \ge 16 \\ 2x - 4y < 8 \end{cases}$

7.

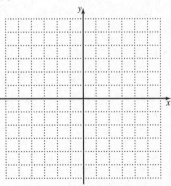

Chapter 10   RATIONAL EXPONENTS, RADICALS, AND COMPLEX NUMBERS

10.1   Radical Expressions and Functions

KEY VOCABULARY

Term	Definition	Example
*n*th root		
Radical function		

KEY PROPERTIES, PROCEDURES, OR STRATEGIES

Evaluating *n*th Roots

Finding the Domain of a Radical Function

Name: Date:

Instructor: Section:

GUIDED EXAMPLES

1. Find each root, if possible.

a) $\sqrt{169}$

Solution

$\sqrt{169} = \boxed{}$ **because (** $)^2 = 169.$

b) $\sqrt{-25}$

Solution

$\sqrt{-25}$ is not a real number because _____ .

_____

c) $\sqrt[3]{-64}$

Solution

$\sqrt[3]{-64} = \boxed{}$ **because (** $)^3 = -64.$

2. Find the domain of $f(x) = \sqrt{3x - 9}$.

Solution

Because $3x - 9$ must be nonnegative, solve $3x - 9 \geq 0$.

$$3x - 9 \geq 0$$

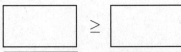

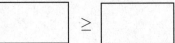

Domain: $\boxed{}$

PRACTICE PROBLEMS

Evaluate each root, if possible.

1. $\sqrt{81}$

1._____

2. $-\sqrt{4}$

2._____

3. $\sqrt[3]{-343}$

3._____

Use a calculator to approximate each root to the nearest thousandth.

4. $\sqrt{13}$

4._____

5. $\sqrt[3]{26}$

5._____

Find the root. Assume that variables represent nonnegative values.

6. $\sqrt{x^{16}}$

6._____

7. $\sqrt{36x^{16}y^{34}}$

7._____

8. $\sqrt[4]{256c^{8}}$

8._____

Find the root. Assume that variables represent any real number.

9. $\sqrt{(x-6)^{2}}$

9._____

10. $\sqrt{(x+6)^{12}}$

11. $\sqrt[3]{(y-2)^9}$

Find the indicated value of the function.

12. $f(x)=\sqrt{3x+88}$; find $f(11)$.

Find the domain.

13. $f(x)=\sqrt{5x+9}$

Solve.

14. Three pieces of lumber are to be connected to form
a right triangle that will be part of a frame for a
roof. If the horizontal piece is 24 feet and the
vertical piece is 10 feet how long must the
connecting piece be?

Chapter 10   RATIONAL EXPONENTS, RADICALS, AND COMPLEX NUMBERS

10.2   Rational Exponents

KEY VOCABULARY

Term	Definition	Example
Rational exponent		

KEY PROPERTIES, PROCEDURES, OR STRATEGIES

Rational Exponents with a Numerator of 1

In the Language of Math	In Your Own Words

General Rule for Rational Exponents

In the Language of Math	In Your Own Words

Negative Rational Exponents

In the Language of Math	In Your Own Words

Rules of Exponents Summary

Multiplying and Dividing Radical Expressions with Different Root Indices

GUIDED EXAMPLE

Use the rules of exponents to simplify. Write the answer with positive exponents. Assume that all variables represent positive values.

$$\left(25x^4 y^6\right)^{3/2}$$

Solution

$$\left(25x^4 y^6\right)^{3/2} = \boxed{}$$ Use $\left(ab\right)^n = a^n \cdot b^n$.

$$= \boxed{}$$ Use $\left(a^m\right)^n = a^{mn}$.

$$= \boxed{}$$ **Multiply the exponents.**

246

PRACTICE PROBLEMS

Rewrite each of the following using radicals; then simplify if possible. Assume that all variables represent nonnegative values.

1. $-25^{1/2}$

1._____

2. $x^{1/5}$

2._____

3. $216^{4/3}$

3._____

4. $256^{-5/4}$

4._____

5. $\left(3y+x\right)^{5/6}$

5._____

Write each of the following in exponential form. Assume that all variables represent positive values.

6. $\sqrt[3]{19}$

6._____

7. $\sqrt[6]{n^5}$

7._____

8. $\dfrac{5x}{\sqrt[5]{z^3}}$

8._____

Use the rules of exponents to simplify. Write the answers with positive exponents. Assume that all variables represent positive values.

9. $a^{7/3}\cdot a^{1/6}$

9._____

10. $\dfrac{6^{9/4}}{6^{7/4}}$

10._____

11. $\left(-6y^{5/6}\right)\left(7y^{-2/3}\right)$

11._____

12. $\left(8x^6y^{18}\right)^{4/3}$

12._____

Represent the following as a radical with a smaller root index. Assume that the variable represents a nonnegative value.

13. $\sqrt[6]{a^4}$

13._____

Perform the indicated operations. Write the result using a radical. Assume that all variables represent positive values.

14. $\sqrt[3]{y}\cdot\sqrt{y}$

14._____

15. $\dfrac{\sqrt[3]{y^5}}{\sqrt[6]{y^5}}$

15._____

Write as a single radical. Assume that all variables represent nonnegative values.

16. $\sqrt[5]{\sqrt[3]{z}}$

16._____

248

Chapter 10   RATIONAL EXPONENTS, RADICALS, AND COMPLEX NUMBERS

10.3   Multiplying, Dividing, and Simplifying Radicals

KEY PROPERTIES, PROCEDURES, OR STRATEGIES

Product Rule for Radicals

In the Language of Math	In Your Own Words

Raising an *n*th Root to the *n*th Power

In the Language of Math	In Your Own Words

Quotient Rule for Radicals

In the Language of Math	In Your Own Words

Simplifying *n*th Roots

Name: Date:

Instructor: Section:

GUIDED EXAMPLES

Simplify. Assume that variables represent positive values.

a) $\sqrt{x^5}$

Solution

$\sqrt{x^5} = $ [   ]   **The greatest number smaller than 5 that is divisible by 2 is _____, so write x^5 as _____.**

$= $ [   ]   **Use the product rule of roots.**

$= $ [   ]   **Simplify.**

b) $-6x^4 y\sqrt{448x^7 y^{19}}$

Solution

$-6x^4 y\sqrt{448x^7 y^{19}}$

$= $ [   ]   **Factor the radicand using the largest possible perfect square factors.**

$= $ [   ]   **Use the product rule of roots.**

$= $ [   ]   **Simplify.**

$= $ [   ]   **Multiply.**

NOTES

PRACTICE PROBLEMS

Find the product and simplify. Assume that variables represent positive values.

1. $\sqrt{10pn^7} \cdot \sqrt{10pn}$

1._____

2. $\sqrt[3]{5a} \cdot \sqrt[3]{17b}$

2._____

3. $\sqrt{\dfrac{x}{5}} \cdot \sqrt{\dfrac{7}{y}}$

3._____

Simplify.

4. $\sqrt{\dfrac{16}{49}}$

4._____

5. $\dfrac{\sqrt{490}}{\sqrt{10}}$

5._____

6. $\dfrac{\sqrt[3]{189}}{\sqrt[3]{7}}$

6._____

Simplify. Assume that variables represent nonnegative values.

7. $\sqrt{40}$

7._____

8. $\sqrt{x^{16}y^{10}}$

8._____

9. $\sqrt[3]{243x^8}$

9._____

Find the product and write the answer in simplest form. Assume that variables represent nonnegative values.

10. $\sqrt{10} \cdot \sqrt{14}$

10._____

11. $a^2\sqrt{a^2b} \cdot b\sqrt{a^3b^4}$

11._____

12. $3\sqrt{2x^5} \cdot 5\sqrt{6x^3}$

12._____

Find the quotient and write the answer in simplest form. Assume that variables represent positive values.

13. $\dfrac{\sqrt{525}}{\sqrt{3}}$

13._____

14. $\dfrac{\sqrt{x^8y^4}}{\sqrt{x^2y^2}}$

14._____

15. $\dfrac{72\sqrt{56x^9y^{14}}}{8\sqrt{7x^6y^5}}$

15._____

Chapter 10   RATIONAL EXPONENTS, RADICALS, AND COMPLEX NUMBERS

10.4   **Adding, Subtracting, and Multiplying Radical Expressions**

KEY VOCABULARY

Term	Definition	Example
Like radicals		

KEY PROPERTIES, PROCEDURES, OR STRATEGIES

Adding Like Radicals

GUIDED EXAMPLE

1.   Add.

$$9\sqrt{32} + 5\sqrt{50}$$

Solution

$$9\sqrt{32} + 5\sqrt{50}$$

= [   ]   **Factor out perfect square factors in each radicand.**

= [   ]   **Use the product rule to separate the radicals.**

= [   ]   **Simplify.**

= [   ]   **Combine like radicals.**

NOTES

GUIDED EXAMPLES

2. Find the product.

   a) $\left(\sqrt{3}+5\right)\left(\sqrt{15}-5\right)$

   Solution

   $\left(\sqrt{3}+5\right)\left(\sqrt{15}-5\right)$

   $=$ [ ]   **Use FOIL.**

   $=$ [ ]   **Use the product rule.**

   $=$ [ ]   **Simplify.**

   b) $\left(5+2\sqrt{3}\right)\left(5-2\sqrt{3}\right)$

   Solution

   $\left(5+2\sqrt{3}\right)\left(5-2\sqrt{3}\right)$

   $=$ [ ]   **Use $(a+b)(a-b)=a^2-b^2$.**

   $=$ [ ]   **Simplify.**

   $=$ [ ]   **Multiply.**

   $=$ [ ]   **Subtract.**

NOTES

Name: _____   Date: _____

Instructor: _____   Section: _____

PRACTICE PROBLEMS

Add or subtract. Assume that variables represent nonnegative values.

1. $8\sqrt{5} - 3\sqrt{5}$

1._____

2. $6b\sqrt[3]{3} - 2b\sqrt[3]{3}$

2._____

3. $\sqrt{80} - \sqrt{45}$

3._____

4. $6\sqrt{125} - 9\sqrt{180}$

4._____

5. $\sqrt{25x^5} - 2\sqrt{36x^5}$

5._____

6. $\sqrt[3]{54x} - \sqrt[3]{2x^4}$

6._____

Use the distributive property. Assume that variables represent nonnegative values.

7. $\sqrt{7}\left(\sqrt{7} + \sqrt{10}\right)$

7._____

8. $2\sqrt{5x}\left(3\sqrt{5x} - 7\sqrt{10x}\right)$

8._____

Multiply. (Use FOIL.)

9. $\left(7\sqrt{7}-4\sqrt{5}\right)\left(8\sqrt{3}+3\sqrt{2}\right)$

9._____

10. $\left(2\sqrt[3]{7}+\sqrt[3]{2}\right)\left(\sqrt[3]{7}-2\sqrt[3]{2}\right)$

10._____

11. $\left(6+\sqrt{10}\right)^2$

11._____

Multiply the conjugates. Assume that variables represent nonnegative values.

12. $\left(\sqrt{3c}+\sqrt{d}\right)\left(\sqrt{3c}-\sqrt{d}\right)$

12._____

13. $\left(\sqrt{2}+4\sqrt{12}\right)\left(\sqrt{2}-4\sqrt{12}\right)$

13._____

Simplify.

14. $\sqrt{15}\cdot\sqrt{5}+15\sqrt{3}$

14._____

15. $6\sqrt{3}\cdot2\sqrt{12}-8\sqrt{18}\cdot\sqrt{6}$

15._____

Chapter 10   RATIONAL EXPONENTS, RADICALS, AND COMPLEX NUMBERS

10.5   Rationalizing Numerators and Denominators of Radical Expressions

KEY PROPERTIES, PROCEDURES, OR STRATEGIES

Rationalizing Denominators

Rationalizing a Denominator Containing a Sum or Difference

GUIDED EXAMPLE

1. Rationalize the denominator. Assume that the variable represents a positive value.

$$\frac{28}{\sqrt{x^3}}$$

Solution

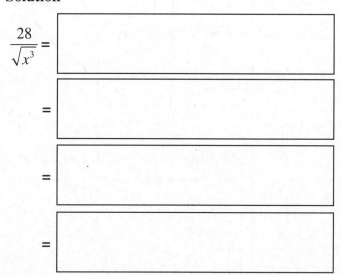

$\dfrac{28}{\sqrt{x^3}} =$ ⬚   **Factor out the largest perfect square in the radicand.**

$=$ ⬚   **Simplify the denominator using the product rule.**

$=$ ⬚   **Multiply by $\dfrac{\sqrt{x}}{\sqrt{x}}$.**

$=$ ⬚   **Simplify.**

GUIDED EXAMPLES

2.   Rationalize the denominator. Assume that variables represent positive values.

a) $\sqrt[3]{\dfrac{3}{25z^2}}$

Solution

$$\sqrt[3]{\frac{3}{25z^2}} = \boxed{}$$

Use the quotient rule to separate the numerator and denominator.

$$= \boxed{}$$

Because $\sqrt[3]{25z^2} \cdot \sqrt[3]{5z} = \sqrt[3]{125z^3} = 5z$, multiply the fraction by $\dfrac{\sqrt[3]{5z}}{\sqrt[3]{5z}}$.

$$= \boxed{}$$

Multiply.

$$= \boxed{}$$

Simplify.

b) $\dfrac{2\sqrt{5}}{\sqrt{5} - 6}$

Solution

$$\frac{2\sqrt{5}}{\sqrt{5} - 6} = \boxed{}$$

The conjugate of $\sqrt{5} - 6$ is $\boxed{}$ so we multiply by $\boxed{}$.

$$= \boxed{}$$

$$= \boxed{}$$

Multiply in the numerator and evaluate the exponents in the denominator.

$$= \boxed{}$$

Simplify.

PRACTICE PROBLEMS

Rationalize the denominator. Assume that variables represent positive values.

1. $\dfrac{5}{\sqrt{28}}$

1._____

2. $\sqrt{\dfrac{4}{125}}$

2._____

3. $\dfrac{\sqrt{3x^2}}{\sqrt{45}}$

3._____

4. $\dfrac{7}{\sqrt{6y}}$

4._____

5. $\dfrac{5}{\sqrt[3]{2}}$

5._____

6. $\sqrt[3]{\dfrac{2}{3}}$

6._____

7. $\dfrac{2}{\sqrt[3]{3x^2}}$

7._____

Rationalize the denominator and simplify. Assume that variables represent positive values.

8. $\dfrac{9}{2+\sqrt{5}}$ **8.**_____

9. $\dfrac{\sqrt{7}}{\sqrt{7}-6}$ **9.**_____

10. $\dfrac{4\sqrt{y}}{\sqrt{y}+6}$ **10.**_____

11. $\dfrac{\sqrt{x}}{\sqrt{x}+\sqrt{y}}$ **11.**_____

Rationalize the numerator. Assume that variables represent positive values.

12. $\dfrac{\sqrt{7x}}{6}$ **12.**_____

13. $\dfrac{\sqrt{11}+2}{7}$ **13.**_____

Chapter 10   RATIONAL EXPONENTS, RADICALS, AND COMPLEX NUMBERS

10.6   Radical Equations and Problem Solving

KEY VOCABULARY

Term	Definition	Example
Radical equation		

KEY PROPERTIES, PROCEDURES, OR STRATEGIES

Power Rule for Solving Equations

Solving Radical Equations

GUIDED EXAMPLE

Solve $\sqrt{y+8} - 2 = \sqrt{y}$.

Solution

$\sqrt{y+8} - 2 = \sqrt{y}$

	Because one of the radicals is isolated, square both sides.
	Simplify.
	Use FOIL on the left-hand side.
	Combine like terms.
	Isolate the remaining radical expression.
	Divide both sides by the coefficient of the radical.
	Square both sides.
	Solve.

Check:

The solution is ____ .

PRACTICE PROBLEMS

Solve.

1. $\sqrt{x} = 7$

1._____

2. $\sqrt[3]{x} = -4$

2._____

3. $\sqrt{x+9} = 11$

3._____

4. $\sqrt{x-3} = 21$

4._____

5. $\sqrt[3]{x+1} = 4$

5._____

6. $\sqrt{y+5} - 5 = 3$

6._____

7. $7 + \sqrt{x-7} = 9$

7._____

8. $\sqrt[4]{y+3} + 2 = -4$

8._____

Solve. Identify any extraneous solutions.

9. $\sqrt{8y-3} = \sqrt{7y+6}$

9._____

10. $\sqrt{2y+4} = \sqrt{2y-6}$

10._____

11. $\sqrt[4]{x+4} = \sqrt[4]{2x}$

11._____

12. $\sqrt{x+31} = x+1$

12._____

13. $\sqrt{y-1} = y-1$

13._____

14. $\sqrt[3]{5x+8} + 4 = 3$

14._____

15. $8 + \sqrt{x} = \sqrt{2x+64}$

15._____

Chapter 10   RATIONAL EXPONENTS, RADICALS, AND COMPLEX NUMBERS

10.7   Complex Numbers

KEY VOCABULARY

Term	Definition	Example
Imaginary unit		
Imaginary number		
Complex number		
Complex conjugate		

KEY PROPERTIES, PROCEDURES, OR STRATEGIES

Rewriting Imaginary Numbers

Name: Date:
Instructor: Section:

GUIDED EXAMPLES

1. Write the imaginary number as a product of a real number and i.
    $\sqrt{-294}$

    Solution

    $\sqrt{-294} =$ [ ]   **Factor out –1 in the radicand.**

    $=$ [ ]   **Use the product rule of square roots.**

    $=$ [ ]   **Factor out the largest possible square from the radicand.**

    $=$ [ ]   **Simplify.**

2. Multiply.
    $(8+i)(2-i)$

    Solution

    $(8+i)(2-i) =$ [ ]   **Use FOIL.**

    $=$ [ ]   **Combine like terms and replace i^2 with –1.**

    $=$ [ ]   **Remove parentheses.**

    $=$ [ ]   **Write in standard form, $a + bi$.**

NOTES

PRACTICE PROBLEMS

Write the imaginary number using i.

1. $\sqrt{-121}$

1._____

2. $\sqrt{-6}$

2._____

3. $\sqrt{-45}$

3._____

Add or subtract.

4. $(6+8i)+(3-7i)$

4._____

5. $(3+4i)-(-5i)$

5._____

6. $(-4+15i)+(-3-9i)+(13+9i)$

6._____

Multiply.

7. $5i(-7+8i)$

7._____

8. $(5+9i)(7-6i)$ 8._____

9. $(-2+4i)^2$ 9._____

Divide and write in standard form.

10. $\dfrac{3}{i}$ 10._____

11. $\dfrac{8}{5+i}$ 11._____

12. $\dfrac{7-4i}{4-i}$ 12._____

Find the powers of i.

13. i^{36} 13._____

14. i^{14} 14._____

15. i^9 15._____

Name: Date:
Instructor: Section:

Chapter 11   QUADRATIC EQUATIONS AND FUNCTIONS

11.1   The Square Root Principle and Completing the Square

KEY PROPERTIES, PROCEDURES, OR STRATEGIES

Square Root Principle

Solving Quadratic Equations by Completing the Square

NOTES

GUIDED EXAMPLE

Solve by completing the square.

$$6x^2 - 7 = -11x$$

Solution

[ ]	**Rewrite in the form $ax^2 + bx = c$.**
[ ]	**Divide both sides by 6.**
[ ]	**Simplify.**
[ ]	**Add $\left(\dfrac{11}{12}\right)^2$ to both sides to complete the square.**
[ ]	**Factor the left side and simplify the right side.**
[ ]	**Use the square root principle.**
[ ]	**Isolate the variable on the left-hand side and simplify the square root.**
[ ]	**Combine like fractions.**

PRACTICE PROBLEMS

Solve and check.

1. $z^2 = 1.96$ 1._____

2. $t^2 = 27$ 2._____

3. $x^2 = -64$ 3._____

Solve and check. Begin by using the addition or multiplication principles of equality to isolate the squared term.

4. $x^2 - 43 = 6$ 4._____

5. $x^2 + 2 = 38$ 5._____

6. $4n^2 = 576$ 6._____

Solve and check. Use the square root principle to eliminate the square.

7. $(3k + 2)^2 = 100$ 7._____

8. $(x - 8)^2 = -81$ 8._____

9. $(0.2x + 5.8)^2 = 0.49$

9._____

Solve the equation by completing the square.

10. $z^2 + 2z + 17 = 0$

10._____

11. $x^2 = -2x + 3$

11._____

12. $x^2 + 3x - 7 = 21$

12._____

Solve the equation by completing the square. Begin by writing the equation in the form $x^2 + bx = c$.

13. $12x^2 - 5 = -17x$

13._____

14. $8x^2 - 15 = 37x$

14._____

15. $3x^2 + 20x - 3 = 0$

15._____

Chapter 11   QUADRATIC EQUATIONS AND FUNCTIONS

11.2   Solving Quadratic Equations Using the Quadratic Formula

KEY VOCABULARY

Term	Definition	Example
Discriminant		

KEY PROPERTIES, PROCEDURES, OR STRATEGIES

Using the Quadratic Formula

Methods for Solving Quadratic Equations

273

Using the Discriminant

GUIDED EXAMPLE
Solve.

$$2x^2 - 7x = 1$$

Solution

| | Write the equation in the form $ax^2 + bx + c = 0$. |

$a = $ _____ $b = $ _____ $c = $ _____   Identify a, b, and c.

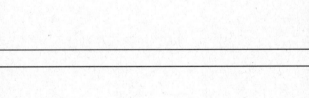

Substitute the appropriate values into the quadratic formula.

Simplify in the radical and the denominator.

PRACTICE PROBLEMS

Solve using the quadratic formula.

1. $x^2 - 3x - 18 = 0$

1._____

2. $x^2 - 4x = -4$

2._____

3. $x^2 - x = -7$

3._____

4. $\dfrac{3}{5}x^2 - x - \dfrac{10}{3} = 0$

4._____

Use the discriminant to determine the number and type of solutions for the equation. If the solutions(s) are real, state whether they are rational or irrational.

5. $x^2 - 8x + 16 = 0$

5._____

6. $\dfrac{1}{4}x^2 - 2x + 4 = 0$

6._____

7. $x^2 - 4x + 9 = 0$

7._____

8. $x^2 - 5x - 5 = 0$ **8.**_____

Find the x- and y-intercepts.

9. $y = x^2 - x - 6$ **9.**_____

10. $y = -x^2 - 6x + 16$ **10.**_____

Translate to a quadratic equation; then solve using the quadratic formula.

11. A positive integer squared plus 4 times its   **11.**_____
consecutive integer is equal to 81. Find the integers.

12. The length of a rectangular air filter is 3 inches less   **12.**_____
than twice the width. Find the length and width of
the filter if the area is 527 square inches.

Solve.

13. If an object is thrown downward with an initial   **13.**_____
velocity of v_0, then the distance it travels is given
by $s = 4.9t^2 + v_0 t$. An object is thrown downward
from an airplane 200 meters from the ground, with
an initial velocity of 8 meters per second. How long
does it take for the object to reach the ground?
Round to the nearest tenth.

276

Name: Date:
Instructor: Section:

Chapter 11   QUADRATIC EQUATIONS AND FUNCTIONS

11.3   Solving Equations That Are Quadratic in Form

KEY VOCABULARY

Term	Definition	Example
Equation quadratic in form		

KEY PROPERTIES, PROCEDURES, OR STRATEGIES

Using Substitution to Solve Equations That Are Quadratic in Form

NOTES

Name: Date:
Instructor: Section:

GUIDED EXAMPLE
Solve.

$$y^{1/3} - y^{1/6} - 12 = 0$$

Solution

$y^{1/3} - y^{1/6} - 12 = 0$ is quadratic in form.

Rewrite in quadratic form.

Substitute u for $y^{1/6}$.

Factor.

Use the zero-factor theorem.

Solve each equation for u.

Substitute $y^{1/6}$ for u.

Raise both sides of each equation to the sixth power.

Simplify.

Check the possible solutions:

Solution(s):

278

PRACTICE PROBLEMS

Solve the equations with rational expressions. Identify any extraneous solutions.

1. $\dfrac{110}{x} - \dfrac{110}{x-5} = -\dfrac{1}{5}$

1._____

2. $\dfrac{1}{x} + \dfrac{1}{x+3} = \dfrac{1}{2}$

2._____

3. $\dfrac{1}{2x-1} - \dfrac{1}{2x+1} = \dfrac{1}{4}$

3._____

4. $6 - t^{-1} - 5t^{-2} = 0$

4._____

Solve the equations with radical expressions. Identify any extraneous solutions.

5. $x - 6\sqrt{x} + 5 = 0$

5._____

6. $x - 7 = \sqrt{x-1}$

6._____

7. $\sqrt{9x-2} - x - 2 = 0$

7._____

8. $\sqrt{3x-2}-\sqrt{4x+1}=-5$ 8._____

Solve using substitution. Identify any extraneous solutions.

9. $x^4-41x^2+400=0$ 9._____

10. $10x^4-9x^2+2=0$ 10._____

11. $(x+3)^2-13(x+3)+42=0$ 11._____

12. $x^{2/3}-2x^{1/3}-3=0$ 12._____

13. $3x^{1/2}+x^{1/4}=2$ 13._____

Solve.

14. Georgia and Benny commute to work daily. Benny   14._____
drives 60 miles and averages 5 miles per hour more
than Georgia. Georgia drives 70 miles, and she is on
the road one-half hour longer than Benny. How fast
does each person drive?

Chapter 11   QUADRATIC EQUATIONS AND FUNCTIONS

11.4   Graphing Quadratic Functions

KEY PROPERTIES, PROCEDURES, OR STRATEGIES

Parabola with Vertex (h, k)

Finding the Vertex of a Quadratic Function in the Form $f(x) = ax^2 + bx + c$

NOTES

Name: Date:
Instructor: Section:

GUIDED EXAMPLE

For the function $f(x) = -(x-3)^2 + 4$,

    a) State whether the graph opens upward or downward.
    b) Find the coordinates of the vertex.
    c) Write the equation of the axis of symmetry.
    d) Graph.

Solution

    a) We see that $f(x) = -(x-3)^2 + 4$ is in the form $f(x) = a(x-h)^2 + k$. The

        parabola opens [upward / downward] because $a =$ ⬚

        which is [positive / negative].

    b) The graph of a function of the form $f(x) = a(x-h)^2 + k$ has vertex (h,k), so the

        vertex of $f(x) = -(x-3)^2 + 4$ is at ⬚ .

    c) For a function of the form $f(x) = a(x-h)^2 + k$, the axis of symmetry is $x = h$.

        The axis of symmetry of $f(x) = -(x-3)^2 + 4$ is $x =$ ⬚ .

    d) To complete the graph, find a few points on either side of the axis of symmetry.

x	y

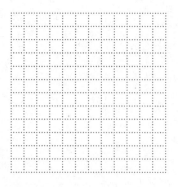

282

PRACTICE PROBLEMS

Find the coordinates of the vertex and write the equation of the axis of symmetry.

1. $h(x) = -6(x-7)^2 + 3$

1._____

_____

2. $f(x) = -x^2 - 6x - 5$

2._____

_____

3. $g(x) = 3x^2 - 18x + 24$

3._____

_____

a.   *State whether the parabola opens upward or downward.*
b.   *Find the coordinates of the vertex.*
c.   *Write the equation of the axis of symmetry.*
d.   *Graph.*

4. $f(x) = -x^2 + 3$

4.a._____

b._____

c._____

d. Graph at the left.

283

Name: Date:
Instructor: Section:

5. $f(x)=\frac{1}{4}(x+2)^2+1$

5.a.________________

b.________________

c.________________

d. Graph at the left.

a. *Find the x- and y-intercepts.*
b. *Write the equation in the form* $f(x)=a(x-h)^2+k$.
c. *State whether the parabola opens upward or downward.*
d. *Find the coordinates of the vertex.*
e. *Write the equation of the axis of symmetry.*
f. *Graph.*
g. *Find the domain and range.*

6. $g(x)=-x^2-2x+3$

6a.________________

b.________________

c.________________

d.________________

e.________________

f. Graph at the left.

g.________________

Name: Date:
Instructor: Section:

7. $h(x) = x^2 + 12x + 36$

7a.________________

b.________________

c.________________

d.________________

e.________________

f. Graph at the left.

g.________________

8. $k(x) = x^2 + 4x + 7$

8a.________________

b.________________

c.________________

d.________________

e.________________

f. Graph at the left.

g.________________

Solve.

9. A toy rocket is launched upward from ground level
    with an initial velocity of 48 meters per second. The
    equation $h = -4.9t^2 + 48t$ describes the height of
    the rocket, where h represents the height in meters
    and t represents the flight time in seconds.
    a. Graph the equation.
    b. What is the maximum height the rocket
        reaches? Round to the nearest thousandth of a
        meter.
    c. How long is the rocket in flight? Round to the
        nearest thousandth of a second.

9a.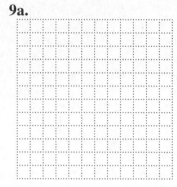

b._____

c._____

10. A carpenter is building a rectangular room with a
    fixed perimeter of 336 feet.
    a. What dimensions would yield the maximum
        area?
    b. What is the maximum area?

10a._____

b._____

11. The amount of nitrogen dioxide, A, in parts per
    million that was present in the air in the city of
    Ovid on a certain day in June is modeled by the
    equation $A = -2t^2 + 40t + 12$, where t is the
    number of hours after 6:00 a.m. Use this equation to
    find the time at which the nitrogen dioxide level
    was at its maximum.

11._____

Chapter 11   QUADRATIC EQUATIONS AND FUNCTIONS

11.5   Solving Nonlinear Inequalities

KEY VOCABULARY

Term	Definition	Example
Quadratic inequality		
Rational inequality		

KEY PROPERTIES, PROCEDURES, OR STRATEGIES

Solving Quadratic Inequalities

NOTES

287

Solving Rational Inequalities

[blank box]

GUIDED EXAMPLE

Solve. Write the solution set using interval notation; then graph the solution set on a number line.

$$x^2 - 15x + 50 < 0$$

Solution

[ ] **Write the related equation.**

[ ] **Factor.**

[ ] **Use the zero-factor theorem and solve.**

These are the *x*-intercepts of the graph of

$$y = x^2 - 15x + 50.$$

Plot the solutions on a number line, which divides the number line into three intervals.

Choose a test number from each interval and substitute that value into $x^2 - 15x + 50 < 0$. If the test number makes the inequality *true*, then all numbers in that interval will solve the inequality. If the test number makes the inequality *false*, then no numbers in that interval will solve the inequality.

Interval			
Test Value			
Result			
True/False			

Solution set: [ ]   Graph:

PRACTICE PROBLEMS

Solve. Write the solution set using interval notation; then graph the solution set on a number line.

1. $(x+3)(x-4)<0$

1._____

⟵————————————⟶

2. $m^2-13m+40>0$

2._____

⟵————————————⟶

3. $r^2-15r+44<0$

3._____

⟵————————————⟶

4. $x^2-2x+1\geq 0$

4._____

⟵————————————⟶

5. $x^2-7x+21<0$

5._____

⟵————————————⟶

6. $(x+5)(3x+12)(2x-10)>0$

6._____

⟵————————————⟶

7. $(x+6)(x+5)(x-2)<0$

7._____

⟵————————————⟶

Solve the rational inequalities. Write the solution set using interval notation; then graph the solution set on a number line.

8. $\dfrac{x-6}{x+5} \le 0$

8._____

9. $\dfrac{8}{x-2} \ge 2$

9._____

10. $\dfrac{b}{b+4} < 2$

10._____

11. $\dfrac{x-7}{x-8} > 3$

11._____

12. $\dfrac{(3x-10)^2}{x+3} > 0$

12._____

Chapter 12   EXPONENTIAL AND LOGARITHMIC FUNCTIONS

12.1 Composite and Inverse Functions

KEY VOCABULARY

Term	Definition	Example
Composition of functions		
Inverse functions		
One-to-one function		

KEY PROPERTIES, PROCEDURES, OR STRATEGIES

Inverse Functions

In the Language of Math	In Your Own Words

Horizontal Line Test for One-to-One Functions

Existence of Inverse Functions

Finding the Inverse Function of a One-to-One Function

Graphs of Inverse Functions

GUIDED EXAMPLE

If $f(x) = x^3 - 9$, find $f^{-1}(x)$.

      Solution

        $f(x) = x^3 - 9$

                                            Replace $f(x)$ with y.

                                            Interchange x and y.

                                            Solve for y.

                                            Replace y with $f^{-1}(x)$.

      Verify by showing that $f\left[f^{-1}(x)\right] = x$ and $f^{-1}\left[f(x)\right] = x$.

Name: Date:

Instructor: Section:

PRACTICE PROBLEMS

Find $(f \circ g)(x)$ and $(g \circ f)(x)$.

1. $f(x) = 2x^2 + 4$, $g(x) = 5x - 3$ 1._____

_____

2. $f(x) = 7x + 5$, $g(x) = \sqrt{x + 8}$ 2._____

_____

3. $f(x) = x^2 + 2$, $g(x) = \sqrt{6 - x}$ 3._____

_____

Determine whether f and g are inverses by determining whether $(f \circ g)(x) = x$ and $(g \circ f)(x) = x$.

4. $f(x) = 10x - 9$, $g(x) = 4 - 2x$ 4._____

5. $f(x) = x^2 + 8$, $x \geq 0$; $g(x) = \sqrt{x - 8}$, $x \geq 8$ 5._____

Determine whether the function is one to one.

6.   6._____

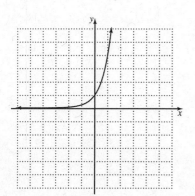

7.

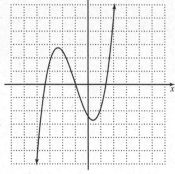

7._____

Find $f^{-1}(x)$ for each of the following one-to-one functions f.

8. $f(x) = 4x + 1$

8._____

9. $f(x) = \dfrac{x-8}{x+9}$

9._____

10. $f(x) = 5x^3 - 6$

10._____

Sketch the graph of the inverse of the following function.

11.

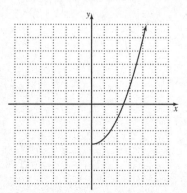

11.

294

Chapter 12   EXPONENTIAL AND LOGARITHMIC FUNCTIONS

12.2 Exponential Functions

KEY VOCABULARY

Term	Definition	Example
Exponential function		

KEY PROPERTIES, PROCEDURES, OR STRATEGIES

One-to-One Property of Exponentials

In the Language of Math	In Your Own Words

Solving Exponential Equations

NOTES

GUIDED EXAMPLES

Solve.

a) $\left(\dfrac{1}{2}\right)^x = 8$

Solution

Write $\dfrac{1}{2}$ as 2^{-1} and 8 as ☐ so that both sides have the same base.

Simplify the left-hand side by applying $\left(a^m\right)^n = a^{mn}$.

Set the exponents equal to each other.

Solve for x.

b) $3^{2x-7} = 81$

Solution

Write 81 as ☐ so that both sides have the same base.

Set the exponents equal to each other.

Solve for x.

NOTES

Name: Date:
Instructor: Section:

PRACTICE PROBLEMS

Graph.

1. $f(x) = 2^{x-3}$

1.
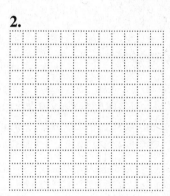

2. $f(x) = 2^x + 3$

2.
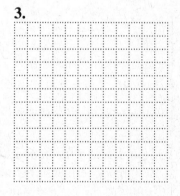

3. $f(x) = \left(\dfrac{1}{6}\right)^x$

3.

Solve each equation.

4. $4^x = 8$

4. _____

5. $5^x = \dfrac{1}{625}$

6. $\left(\dfrac{3}{4}\right)^x = \dfrac{64}{27}$

Use the formula $A = P\left(1 + \dfrac{r}{n}\right)^{nt}$.

7. If \$19,000 is deposited into an account paying 6% interest compounded quarterly, how much will be in the account after 7 years?

A culture of a bacteria doubles in size every 40 minutes. If A_0 *is the initial amount and t is the number of minutes passed, the amount present is A, where* $A = A_0 (2)^{t/40}$.

8. If a culture of the bacteria began with 2000 cells, how many cells will be present after 20 minutes? Round to the nearest whole number.

Chapter 12   EXPONENTIAL AND LOGARITHMIC FUNCTIONS

12.3 Logarithmic Functions

KEY VOCABULARY

Term	Definition	Example
Logarithm		

KEY PROPERTIES, PROCEDURES, OR STRATEGIES

Solving Logarithmic Equations

Logarithmic Properties

In the Language of Math	In Your Own Words

Graphing Logarithmic Functions

Name: Date:
Instructor: Section:

GUIDED EXAMPLES

Solve.

a) $\log_x 81 = 4$

Solution

$\log_x 81 = 4$

Write in exponential form.

Find the positive and negative fourth roots of 81.

The base must be positive.

b) $\log_3 x + 3 = 7$

Solution

$\log_3 x + 3 = 7$

Subtract 3 from both sides.

Write in exponential form.

Simplify.

NOTES

PRACTICE PROBLEMS

Write in logarithmic form.

1. $3^5 = 243$

1. _____

2. $64^{1/3} = 4$

2. _____

3. $\left(\dfrac{1}{5}\right)^{-3} = 125$

3. _____

Write in exponential form.

4. $\log_5 625 = 4$

4. _____

5. $\log_{10} \dfrac{1}{100,000} = -5$

5. _____

6. $\log_2 \dfrac{1}{4} = -2$

6. _____

Solve.

7. $\log_2 x = -1$

7. _____

8. $\log_4 256 = x$

8. _____

9. $\log_2 \dfrac{1}{8} = x$

9. _____

301

10. $\log_x \dfrac{1}{27} = -3$

10._____

11. $\log_4 x + 5 = 8$

11._____

12. $3\log_b 64 = 9$

12._____

Graph.

13. $f(x) = \log_6 x$

13.

Solve.

14. The percent of adult length attained by a 2- to 18-month old reptile of a certain species can be approximated by $f(x) = 38 + 30\log_3 (x - 11)$, where x is the reptile's age in months and $f(x)$ is the percent. At an age of 14 months, what percent of its adult length has the reptile reached?

14._____

Chapter 12   EXPONENTIAL AND LOGARITHMIC FUNCTIONS

12.4 Properties of Logarithms

KEY PROPERTIES, PROCEDURES, OR STRATEGIES

Inverse Properties of Logarithms

In the Language of Math	In Your Own Words

Further Properties of Logarithms

GUIDED EXAMPLES

1. Write the expression as the sum or difference of logarithms.

$$\log_4\left(\frac{y}{3z}\right)$$

Solution

$$\log_4\left(\frac{y}{3z}\right)$$

Use the quotient rule.

Use the product rule. Use parentheses if necessary.

Remove the parentheses.

2. Write the expression as a single logarithm. Leave the answer in simplest form without negative or fractional exponents.

$$3\log_a 4 - 2\log_a 3$$

Solution

$$3\log_a 4 - 2\log_a 3$$

Use the power rule.

Use the quotient rule.

Simplify.

NOTES

PRACTICE PROBLEMS

Find the value.

1. $12^{\log_{12} f}$ 1._____

2. $\log_q q^{18}$ 2._____

Use the product rule to write the expression as a sum of logarithms.

3. $\log_d (aqr)$ 3._____

4. $\log_b y(y-4)$ 4._____

Use the product rule to write the expression as a single logarithm.

5. $\log_3 H + \log_3 Z$ 5._____

6. $\log_3 (x+6) + \log_3 (x+10)$ 6._____

Use the quotient rule to write the expression as a difference of logarithms. Leave your answers in simplest form.

7. $\log_6 \dfrac{13}{7}$ 7._____

8. $\log_d \dfrac{d}{k}$ 8._____

Use the quotient rule to write the expression as a single logarithm. Leave your answers in simplest form.

9. $\log_c 35 - \log_c 5$

9._____

10. $\log_a \left(x^2 - 16 \right) - \log_a \left(x - 4 \right)$

10._____

Use the power rule to write the expression as a multiple of a logarithm.

11. $\log_N t^{15}$

11._____

12. $\log_7 \sqrt[3]{b}$

12._____

Use the power rule to write the expression as a logarithm of a quantity to a power. Leave your answers in simplest form without negative or fractional exponents.

13. $2\log_6 7$

13._____

14. $-3\log_2 a$

14._____

Write the expression as the sum or difference of multiples of logarithms.

15._____

15. $\log_b \dfrac{xy^7}{z^6}$

Write the expression as a single logarithm. Leave your answer in simplest form without negative or fractional exponents.

16. $4\log_a x + 7\log_a z$

16._____

306

Chapter 12 EXPONENTIAL AND LOGARITHMIC FUNCTIONS

12.5 Common and Natural Logarithms

KEY VOCABULARY

Term	Definition	Example
Common logarithms		
Natural logarithms		

NOTES

GUIDED EXAMPLE

The magnitude of an earthquake is given by the Richter scale, whose formula is $R = \log \dfrac{I}{I_0}$, where I is the intensity of the earthquake and I_0 is the intensity of a minimal earthquake and is used for comparison purposes. The 1989 Loma Prieta earthquake had a magnitude of 7.1, and the 1994 Northridge earthquake had a magnitude of 6.7. Compare the intensity of the two earthquakes.

Solution

Understand We are to compare the intensities of two earthquakes, one with a magnitude of 7.1 and the other with a magnitude of 6.7.

Plan   In $R = \log \dfrac{I}{I_0}$, replace R with 7.1 and express I in terms of I_0. Repeat this process for $R = 6.7$.

(continued on the next page)

Name: Date:
Instructor: Section:

GUIDED EXAMPLE *(continued)*

Execute

1989 earthquake: 1994 earthquake:

$$R = \log \frac{I}{I_0}$$ $$R = \log \frac{I}{I_0}$$

Replace R with the
magnitude of each
earthquake.

$$7.1 = \log \frac{I}{I_0}$$ $$6.7 = \log \frac{I}{I_0}$$

[ ]   **Write each equation in**   [ ]
                             exponential form.

[ ]   **Multiply both sides of**   [ ]
                             each equation by I_0.

[ ]   **Compare the intensities by writing the**
             expression for the intensity of the 1989
             earthquake divided by expression for
             the intensity of the 1994 earthquake.

[ ]   **Divide out the common factor of I_0. Use**
             properties of exponents to write the
             exponential expressions with a single
             base.

[ ]   **Use a calculator to evaluate the**
             exponential expression. The result is
             how many times stronger the 1989
             earthquake was compared to the 1994
             earthquake.

Answer   The 1989 earthquake was _____ times stronger than the 1994
earthquake.

Check   Verify your calculations and make sure your answer makes sense.

308

PRACTICE PROBLEMS

Use a calculator to approximate each logarithm to four decimal places.

1. $\log 93$ 1._____

2. $\log\left(9.15 \times 10^{-3}\right)$ 2._____

3. $\log\left(5.51 \times 10^{4}\right)$ 3._____

4. $\ln 65.3$ 4._____

5. $\ln\left(4.54 \times e^{7}\right)$ 5._____

Find the exact value of each logarithm using $\log_b b^x = x$.

6. $\log 0.0001$ 6._____

7. $\log 1$ 7._____

8. $\log \sqrt[5]{10}$ 8._____

9. $\ln\left(e^{-1}\right)$ 9._____

Solve.

10. The intensity I of a particular sound has a decibel
   reading of $d = 10 \log \dfrac{I}{I_0}$. What is the decibel
   reading of this sound if its intensity is 10^{-7} and
   $I_0 = 10^{-12}$ watts/m^2?

   10._____

The pH of a substance determines whether it is a base (pH > 7) or an acid (pH < 7). To find the pH of a solution, we use the formula $pH = -\log\left[H_3O^+\right]$, *where* $\left[H_3O^+\right]$ *is the hydronium ion concentration in moles per liter.*

11. Find the pH of a certain agricultural product with a
   hydronium ion concentration of 3.8×10^{-4} moles
   per liter. Round to the nearest tenth.

   11._____

12. The pH of a fruit juice is 3.6. Find the hydronium
   ion concentration of the juice.

   12._____

Chapter 12   EXPONENTIAL AND LOGARITHMIC FUNCTIONS

12.6 Exponential and Logarithmic Equations with Applications

KEY PROPERTIES, PROCEDURES, OR STRATEGIES

Properties for Solving Exponential and Logarithmic Equations

Solving Equations Containing Logarithms

Change-of-Base Formula

In the Language of Math	In Your Own Words

Name: Date:
Instructor: Section:

GUIDED EXAMPLE
Solve.

$$\log_4 x + \log_4 (x-6) = 2$$

Solution

$$\log_4 x + \log_4 (x-6) = 2$$

	Use $\log_b xy = \log_b x + \log_b y$ to simplify the left side.
	The equation is in the form $\log_b x = y$; so write it in the exponential form, $b^y = x$.
	Evaluate the exponential expression and use the distributive property.
	Write the equation in the form $ax^2 + bx + c = 0$.
	Factor.
	Set each factor equal to 0.
	Solve each equation to find possible solutions.
	Check each answer. Remember that logarithms are defined for positive numbers only.

Solution(s): _____

312

PRACTICE PROBLEMS

Solve. Round your answers to four decimal places.

1. $2^x = 6$ **1.** _____

2. $4^{-x+4} = 85$ **2.** _____

3. $e^{0.006x} = 29$ **3.** _____

4. $e^{-0.596x} = 5$ **4.** _____

Solve. Give exact answers.

5. $\log_3 (3x - 6) = 3$ **5.** _____

6. $\log x + \log (x + 15) = 2$ **6.** _____

7. $\log_3 (x) + \log_3 (x - 1) = \log_3 2$ **7.** _____

8. $\log 8x - \log (2x - 3) = \log 9$ **8.** _____

9. $\log_6 (4t + 7) - \log_6 t = \log_6 5$ **9.** _____

Solve.

10. How long will it take for \$28,000 to grow to   10._____
\$30,616.41 at 6% interest compounded quarterly?
Round to the nearest tenth of a year.

11. Suppose that \$70,000 is invested at 5% interest.   11a._____
a. Find the amount of money in the account after 7
years if the interest is compounded annually.
b. Find the amount of money in the account after 7   b. _____
years if the interest is compounded continuously.

12. The population of a city was 98 thousand in 1992. The   12.a._____
exponential growth rate was 1.8% per year. The population
t years after 1992 is given by $P = 98,000e^{0.018t}$.
a. Estimate the population in 2006 to the nearest   b. _____
thousand.
b. Find the time at which the population reached
110 thousand. Round to the nearest year.

Use the change-of-base formula to find the logarithm. Round your answer to four decimal places.

13. $\log_7 10$ 13._____

Use $P = 95 - 30\log_2 x$ and the change-of-base formula to find the percentage, P, of the lecture retained after the given number of days. Round your answer to the nearest whole percent.

14. 6 days   14._____

Chapter 13   CONIC SECTIONS

13.1 Parabolas and Circles

KEY VOCABULARY

Term	Definition	Example
Conic section		
Circle		
Radius		

KEY PROPERTIES, PROCEDURES, OR STRATEGIES

Equations of Parabolas Opening Left or Right

Distance Formula

In the Language of Math	In Your Own Words

315

Name: Date:

Instructor: Section:

Midpoint Formula

Standard Form of the Equation of a Circle

GUIDED EXAMPLE

For $x = -y^2 + 2y + 1$, determine whether the graph opens left or right, find the vertex and axis of symmetry, and draw the graph.

    Solution

    This parabola opens [left / right] because $a = $ _____, which is [positive / negative].

    Write the equation in the form $x = a(y-k)^2 + h$.

$x = -y^2 + 2y + 1$ **Original equation**

         Subtract 1 from both sides.

         Factor out –1 from the right-hand side.

         Complete the square.

         Simplify the left side and factor the right side.

         Isolate the variable x.

Vertex: _____   Axis of symmetry: _____

To complete the graph, let y equal values on each side of the axis of symmetry and find x.

x	y

PRACTICE PROBLEMS

Find the direction the parabola opens, the coordinates of the vertex, and the equation of the axis of symmetry and draw the graph.

1. $y = -x^2 - 4x - 3$

1._____

        _____

        _____

2. $x = y^2 + 6y + 3$

2._____

        _____

        _____

Find the distance and midpoint between the two points.

3. $(-10, -2)$ and $(10, -17)$

3._____

317

4. $(5,1)$ and $(2,-6)$ 4._____

Find the center and radius and draw the graph.

5. $(x+2)^2 + (y+1)^2 = 9$ 5._____

                                              _____

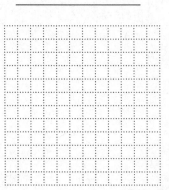

6. $x^2 + y^2 - 6x + 4y = 23$ 6._____

                                              _____

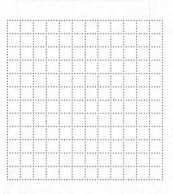

The center and radius of a circle are given. Write the equation of the circle in standard form.

7. Center: $(9,4)$; radius: 2   7._____

8. Center: $(-1,3)$; radius: $\sqrt{6}$ 8._____

318

Chapter 13   CONIC SECTIONS

13.2 Ellipses and Hyperbolas

KEY VOCABULARY

Term	Definition	Example
Ellipse		
Hyperbola		

KEY PROPERTIES, PROCEDURES, OR STRATEGIES

Equation of an Ellipse Centered at (0,0)

General Equation for an Ellipse

Equations of Hyperbolas in Standard Form

Graphing a Hyperbola in Standard Form

GUIDED EXAMPLE

Graph the ellipse.

$$\frac{(x-1)^2}{25}+\frac{(y+5)^2}{4}=1$$

Solution

The center of the ellipse is $(h,k)=$ ☐ .

Also, $a=$ ☐ and $b=$ ☐ so the

ellipse passes through the following four points:

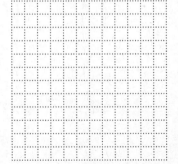

PRACTICE PROBLEMS

Graph each ellipse. Label the x- and y-intercepts.

1. $\dfrac{x^2}{9} + \dfrac{y^2}{4} = 1$

1.

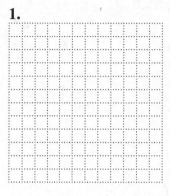

2. $16x^2 + 4y^2 = 64$

2.

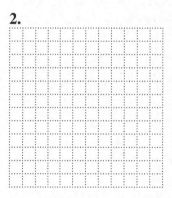

Graph the ellipse. Label the center and the points directly above, below, to the left, and to the right of the center.

3. $\dfrac{(x-3)^2}{16} + \dfrac{(y+2)^2}{9} = 1$

3.

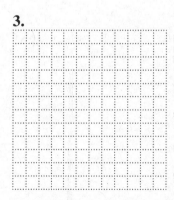

Name: Date:

Instructor: Section:

Graph each hyperbola. Also show the fundamental rectangle with its corner points labeled, the asymptotes, and the intercepts.

4. $\dfrac{x^2}{4} - \dfrac{y^2}{16} = 1$

4.

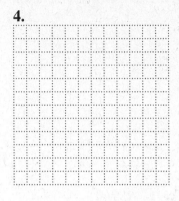

5. $16x^2 - y^2 = 64$

5.

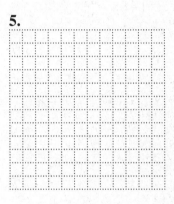

6. $25y^2 - 9x^2 = 225$

6.

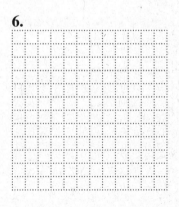

Chapter 13   CONIC SECTIONS

13.3 Nonlinear Systems of Equations

KEY VOCABULARY

Term	Definition	Example
Nonlinear system of equations		

GUIDED EXAMPLE

Solve the system by substitution.

$$\begin{cases} y = x^2 + 7x + 10 \\ x + y = 3 \end{cases}$$

Solution

	Solve $x + y = 3$ for y.
	Substitute this expression for y in the first equation.
	Write the equation in the form $ax^2 + bx + c = 0$.
	Factor the left-hand side.
	Use the zero-factor theorem.
	Solve each equation.
	To find y, substitute these solutions for x in either of the original equations.

Solution(s):

Name: Date:

Instructor: Section:

GUIDED EXAMPLE

Solve the system by elimination.

$$\begin{cases} x^2 + y^2 = 106 \\ x^2 - y^2 = 56 \end{cases}$$

Solution

The graphs of these equations are a circle and a hyperbola.

$$x^2 + y^2 = 106$$
$$\underline{x^2 - y^2 = 56}$$

To eliminate y^2, add the equations.

Divide both sides by 2.

Find the square roots.

To find y, substitute the values for x in one of the original equations.

Solution(s):

NOTES

Name: Date:
Instructor: Section:

PRACTICE PROBLEMS

Solve.

1. $\begin{cases} x^2 + y = 27 \\ 3x - y = 1 \end{cases}$

   1._____

2. $\begin{cases} y = x^2 + 3x + 9 \\ x + y = 6 \end{cases}$

   2._____

3. $\begin{cases} x^2 + y^2 = 100 \\ y - x = 2 \end{cases}$

   3._____

4. $\begin{cases} y = x^2 - 24 \\ y = 2x \end{cases}$

   4._____

5. $\begin{cases} x^2 + y^2 = 100 \\ x^2 - y^2 = -28 \end{cases}$

   5._____

Name: Date:
Instructor: Section:

6. $\begin{cases} x^2 + y^2 = 20 \\ 3x^2 - y^2 = 44 \end{cases}$

6._____

7. $\begin{cases} x^2 + 225y^2 = 25 \\ x = y^2 + 5 \end{cases}$

7._____

8. $\begin{cases} xy = 100 \\ 16x^2 + y^2 = 800 \end{cases}$

8._____

9. $\begin{cases} y = x^2 + 2x - 3 \\ y = -x^2 + 4x + 1 \end{cases}$

9._____

10. $\begin{cases} -2x^2 + 5y^2 = 75 \\ 4x^2 + 3y^2 = 175 \end{cases}$

10._____

326

Chapter 13   CONIC SECTIONS

13.4 Nonlinear Inequalities and Systems of Inequalities

KEY PROPERTIES, PROCEDURES, OR STRATEGIES

Graphing Nonlinear Inequalities

GUIDED EXAMPLE
Graph the inequality.

$$\frac{x^2}{16} + \frac{y^2}{4} \geq 1$$

Solution

The related equation is an ellipse with center $(0,0)$ and $a = 4$ and $b = 2$. Because the

inequality is $\geq$, all ordered pairs on the ellipse are in the solution set, and we draw it

with a [solid / dashed] curve. Choose a test point to determine which region to shade.

Test point:

The statement is [true / false]. Shade the

appropriate region of the graph.

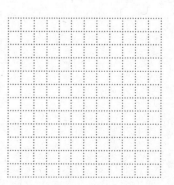

Name: Date:

Instructor: Section:

GUIDED EXAMPLE

Graph the solution set of the system of inequalities.

$$\begin{cases} \dfrac{x^2}{9} - \dfrac{y^2}{9} > 1 \\ y > 1 \end{cases}$$

Solution

Graph each inequality separately. For each inequality, begin by graphing the related equation, and determine whether the graph should be solid or dashed. Then use a test point to determine which region to shade.

$$\frac{x^2}{9} - \frac{y^2}{9} > 1$$

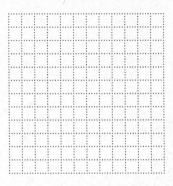

$$y > 1$$

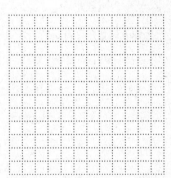

The solution set for the system contains all ordered pairs in the region where the two graphs overlap.

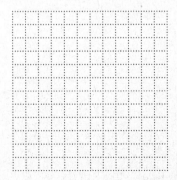

328

PRACTICE PROBLEMS

Graph each inequality.

1. $x^2 + y^2 < 25$

1.

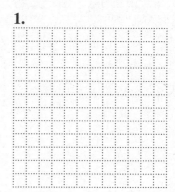

2. $y \geq x^2 - 5$

2.

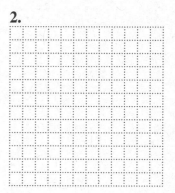

3. $\dfrac{x^2}{9} + \dfrac{y^2}{4} < 1$

3.

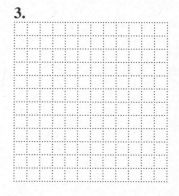

4. $y^2 - \dfrac{x^2}{4} \geq 1$

4.

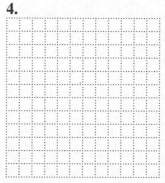

329

Name: Date:

Instructor: Section:

Graph the solution set of each system of inequalities.

5. $\begin{cases} y \le x^2 \\ 3x - 4y < 12 \end{cases}$

5.

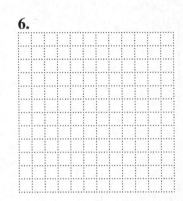

6. $\begin{cases} -x - 3y > 3 \\ x^2 + y^2 < 25 \end{cases}$

6.

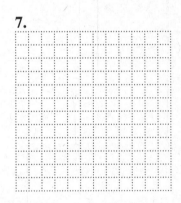

7. $\begin{cases} x^2 + y^2 \ge 16 \\ x^2 + y^2 < 36 \end{cases}$

7.

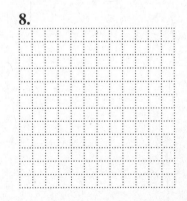

8. $\begin{cases} y \ge x^2 - 4 \\ y \le -x^2 + 2 \end{cases}$

8.

APPENDICES

Appendix A   Arithmetic Sequences and Series

KEY VOCABULARY

Term	Definition	Example
Sequence		
Finite sequence		
Infinite sequence		
Arithmetic sequence		
Common difference of an arithmetic sequence		
Series		
Arithmetic series		

KEY PROPERTIES, PROCEDURES, OR STRATEGIES

nth Term of an Arithmetic Sequence

Partial Sum, S_n, of an Arithmetic Series

GUIDED EXAMPLES

1. Find the 16th term and an expression for the nth term of an arithmetic sequence in which $a_1 = -5$ and $d = 7$.

   Solution

   $a_n = a_1 + (n-1)d$

   Substitute 16 for n, –5 for a_1, and 7 for d.

   Evaluate

   To find an expression for the nth term, substitute for a_1 and d.

   Simplify.

2. The first term of an arithmetic sequence is 14 and the 15th term (a_{15}) is -28. Find the common difference and the first four terms of the sequence.

   Solution

   Use the fact that we know a_{15} and the formula for the nth term to find d.

   $a_n = a_1 + (n-1)d$

   Substitute 15 for n and 14 for a_1.

   Substitute –28 for a_{15} and solve.

   The first four terms are _____ .

Name: Date:
Instructor: Section:

PRACTICE PROBLEMS

Write the first four terms of the sequence and the indicated term.

1. $a_n = n^2 - 1$, 10th term

1._____

2. $a_n = \dfrac{(-1)^n}{n+1}$, 15th term

2._____

Find the common difference, d, for each arithmetic sequence.

3. 8, 19, 30, 41

3._____

4. $-37, -34, -31, -28$

4._____

Find the indicated term and an expression for the nth term of the given arithmetic sequence.

5. a_6 if $a_1 = 7$ and $d = 2$

5._____

6. a_{11} if $a_1 = -10$ and $d = -4$

6._____

Write the first four terms of the arithmetic sequence with the given characteristics.

7. $a_1 = -18$, $d = 6$

7._____

8. $a_1 = 14,\ a_8 = -42$ **8.**_____

Write the first four terms of the arithmetic sequence with the given d and a_n.

9. $d = -5,\ a_6 = 13$ **9.**_____

10. $d = 3,\ a_{12} = 1$ **10.**_____

Write the series and find the sum.

11. $\displaystyle\sum_{i=1}^{8} (i-3)$ **11.**_____

12. $\displaystyle\sum_{i=1}^{5} (3i+4)$ **12.**_____

Find the given S_n for the arithmetic series.

13. If $a_1 = 9$ and $d = 6$, find S_{20}.   **13.**_____

14. $12 + 5 + (-2) + (-9) + \cdots$. Find S_{10}.   **14.**_____

334

APPENDICES

Appendix B   Geometric Sequences and Series

KEY VOCABULARY

Term	Definition	Example
Geometric sequence		
Geometric series		

KEY PROPERTIES, PROCEDURES, OR STRATEGIES

nth Term of a Geometric Sequence

Partial Sum, S_n, of a Geometric Series

Sum of an Infinite Geometric Series

335

GUIDED EXAMPLES

1. Find the sum of the first 8 terms of the geometric series: $3-12+48-192+\cdots$

   Solution

   We first find r: $r = \dfrac{\boxed{}}{\boxed{}} = \boxed{}$

   $S_n = \dfrac{a\left(1-r^n\right)}{1-r}$ **Formula for S_n**

   $\boxed{}$ **Substitute 8 for n, 3 for a, and the value we found for r. Be sure to use parentheses where necessary.**

   $\boxed{}$ **Evaluate.**

2. Find the sum of the infinite geometric series, if possible: $18+6+2+\dfrac{2}{3}+\cdots$

   Solution

   We know that $a = \boxed{}$ and $r = \boxed{}$. Because $|r| < 1$, the sum exists.

   $S_\infty = \dfrac{a}{1-r}$

   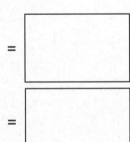

   $= \boxed{}$ **Substitute the values for a and r.**

   $= \boxed{}$ **Evaluate.**

   NOTES

Name: Date:
Instructor: Section:

PRACTICE PROBLEMS

a. Find the common ratio, r, for the given geometric sequence.
b. Find the indicated term.
c. Find an expression for the general term, a_n.

1. $4, 12, 36, 108, \cdots$; 10th term

1a._____

b._____

c._____

2. $12, -6, 3, -\dfrac{3}{2}, \cdots$; 6th term

2a._____

b._____

c._____

a. Write the first five terms of the geometric sequence satisfying the given conditions.
b. Find the indicated term.

3. $a = -1, \ r = -4; \ 10^{\text{th}}$ term

3a._____

b._____

4. $a = 125, \ r = \dfrac{1}{5}; \ 8^{\text{th}}$ term

4a._____

b._____

a. Use the formula for the nth term to find r.
b. Write the first four terms of the geometric sequence.

5. $a = 8, \ a_5 = 648, \ r > 0$

5a._____

b._____

6. $a = 144, \ a_6 = \dfrac{9}{2}$

6a._____

b._____

337

Find the sum of the first n terms of each geometric series for the given value of n.

7. $-4-8-16-32-\cdots$, $n=9$ **7.**_____

8. $5-10+20-40+\cdots$, $n=8$ **8.**_____

Find the sum of the infinite geometric series, if possible. If it is not possible, explain why.

9. $128+32+8+\cdots$ **9.**_____

10. $-2+4-8+\cdots$ **10.**_____

Write each repeating decimal as a fraction.

11. $0.\overline{8}$ **11.**_____

12. $0.\overline{31}$ **12.**_____

Solve.

13. A new laptop computer costs $2500 and it   **13.**_____
depreciates by 20% each year. What will the laptop
be worth in 5 years?

APPENDICES

Appendix C   The Binomial Theorem

KEY VOCABULARY

Term	Definition	Example
Factorial notation		
Binomial coefficient		
Binomial theorem		

KEY PROPERTIES, PROCEDURES, OR STRATEGIES

Finding the $(m + 1)$st Term of a Binomial Expansion

GUIDED EXAMPLES

1. Evaluate the binomial coefficient $\binom{9}{5}$.

Solution

$$\binom{9}{5} = \boxed{}$$

Substitute 9 for n and 5 for r in $\dfrac{n!}{r!(n-r)!}$.

$$= \boxed{}$$

Simplify.

$$= \boxed{}$$

Rewrite 9! as $9 \cdot 8 \cdot 7 \cdot 6 \cdot 5!$

$$= \boxed{}$$

Evaluate.

2. Expand $(x+y)^4$ using the binomial theorem.

Solution

$(x+y)^4$

$$= \left(\right) x^{\square} + \left(\right) x^{\square} y + \left(\right) x^{\square} y^{\square} + \left(\right) x y^{\square} + \left(\right) y^{\square}$$

$$= \boxed{}$$

$$= \boxed{}$$

NOTES

340

PRACTICE PROBLEMS

Evaluate each expression.

1. $(5!)(4!)$

1._____

2. $\dfrac{12!}{8!}$

2._____

3. $\dfrac{6!}{4!(6-4)!}$

3._____

Evaluate each binomial coefficient.

4. $\dbinom{8}{3}$

4._____

5. $\dbinom{6}{1}$

5._____

6. $\dbinom{7}{4}$

6._____

Use the binomial theorem to expand each of the following.

7. $(x+y)^4$

8. $(a+2b)^5$

8._____

9. $(3c-d)^6$

9._____

10. $(4a+b)^8$

10._____

Find the indicated term of each binomial expansion.

11. $(x-y)^6$, fourth term

11._____

12. $(2a+b)^5$, third term

12._____

APPENDICES

Appendix D   Synthetic Division

KEY PROPERTIES, PROCEDURES, OR STRATEGIES

The Remainder Theorem

In the Language of Math	In Your Own Words

GUIDED EXAMPLE

Divide using synthetic division.

$$\frac{4y^3 + 14y^2 + 8y - 11}{y + 2}$$

Solution

First, we need the divisor to be in the form $y - c$. So we rewrite $y + 2$ as $y - (-2)$,

and so we see that $c = $ [＿＿＿].

We then set up the synthetic division with c and the coefficients of the dividend.

$$-2 \rfloor \quad 4 \qquad 14 \qquad 8 \qquad -11$$

The first step is to bring down the leading coefficient.
Multiply it by the value c and write the result underneath the next coefficient above the horizontal line.
Add the numbers in that column, and write the result in the same column below the horizontal line.
Multiply the result by the value c and write the result underneath the next coefficient above the horizontal line. Continue this process to the last column. The final result is the remainder.

Answer: [＿＿＿＿＿]

The degree of the quotient is one less than the degree of the dividend.

Name: _____   Date: _____

Instructor: _____   Section: _____

GUIDED EXAMPLE

Divide using synthetic division.

$$\frac{2x^3 - 5x^2 + 3x - 2}{x - 2}$$

Solution

The divisor is already in the form $y - c$. So we see that $c = $ ⬚.

We then set up the synthetic division with c and the coefficients of the dividend.

```
 2|    2    −5     3    −2
           [ ]   [ ]   [ ]
      _____
     [ ]   [ ]   [ ]   [ ]
```

The first step is to bring down the leading coefficient.
Multiply it by the value c and write the result underneath the next coefficient above the horizontal line.
Add the numbers in that column, and write the result in the same column below the horizontal line.
Multiply the result by the value c and write the result underneath the next coefficient above the horizontal line. Continue this process to the last column. The final result is the remainder.

Answer: ⬚

The degree of the quotient is one less than the degree of the dividend.

NOTES

344

PRACTICE PROBLEMS

Divide using synthetic division.

1. $\dfrac{x^2 - 7x + 12}{x - 3}$

          1._____

2. $\dfrac{3x^2 - x - 2}{x - 4}$

          2._____

3. $\dfrac{2r^3 + 15r^2 + 24r - 5}{r + 5}$

          3._____

4. $\dfrac{4w^3 + 20w^2 + 13w - 15}{w + 4}$

          4._____

5. $\dfrac{2x^3 - 34x - 14}{x + 4}$

          5._____

6. $\dfrac{y^3 + 3}{y - 3}$

6._____

7. $\dfrac{s^4 - 625}{s - 5}$

7._____

8. $\dfrac{2x^3 - 3x^2 + 4x - 3}{x - 1}$

8._____

9. $\dfrac{x^3 - 9x^2 + 11}{x + 1}$

9._____

10. $\dfrac{x^4 - x^3 - 13x^2 + 5x - 3}{x - 4}$

10._____

APPENDICES

Appendix E   Mean, Median, and Mode

KEY VOCABULARY

Term	Definition	Example
Statistic		
Mean or arithmetic average		
Median		
Mode		

KEY PROPERTIES, PROCEDURES, OR STRATEGIES

Finding the Mean, or Arithmetic Average, of a Given Set of Numbers

Finding the Median of a Set of Numbers

Finding the Mode of a Set of Numbers

347

Name: Date:
Instructor: Section:

GUIDED EXAMPLES

The following numbers are the total number of viewers (in millions) for the first six episodes of a top-rated television show.

    19.3   18.6   17.2   19.8   20.3   18.6

a) Find the mean.

    Solution

    Understand: We must calculate the arithmetic average, or mean, number of viewers. We are given the number of viewers for the first six episodes.

    Plan: Divide the sum of the numbers by the number of numbers in the list.

    Execute:

    Answer: The mean number of viewers for the show is _____ million viewers.

    Check: Verify the calculations by inverse operations.

b) Find the median.

    Solution

    Understand: We must find the median number of viewers for the show.

    Plan: Arrange the numbers in order from least to greatest. Because there is an even number of numbers, the median will be the mean of the middle two numbers.

    Execute:

    Answer: The median number of viewers is _____ million viewers.

    Check: Verify the calculations by inverse operations.

c) Find the mode.

    Solution

    Understand: We must find the mode of the set of numbers. The mode is the number that occurs most frequently.

    Plan: Order the data and count then number of repetitions of each number. The number with the most repetitions is the mode.

    Execute:

    Answer: The mode is _____ million viewers.

Name: Date:
Instructor: Section:

PRACTICE PROBLEMS

Solve.

1. Following is a list of the salaries of the employees
    at the Hemms Etc. clothing store. Find the mean,
    median, and mode of the salaries.

$26,100	$16,700
$18,700	$11,000
$18,700	$21,100

    1._____

    _____

    _____

2. Following is a list of hourly wages for employees at
    a call center. Find the mean, median, and mode of
    the wages.

$18.00	$21.50	$17.30
$17.00	$19.80	$19.40
$20.25	$19.40	$18.60

    2._____

    _____

    _____

3. Find the mean, median, and mode for the daily auto
    sales at a local car dealership for a week.
    12, 14, 25, 10, 25, 15, 11

    3._____

    _____

    _____

4. Find the mean, median, and mode for the number of
    customers at an ice cream stand for a week.
    112, 64, 52, 71, 85, 102, 130

    4._____

    _____

    _____

349

Name: Date:
Instructor: Section:

5. Find the mean, median, and mode for the number of
    hours that different students spent studying for an
    exam.
    1.7, 4.8, 6.2, 7.9, 7.9

5._____

_____

_____

6. The Buffalo Bills scored the following numbers of
    points during 2009 regular season games. Find the
    mean, median, and mode of the scores.

    | 24 | 33 | 7 | 10 | 3 | 16 |
    | 20 | 10 | 17 | 15 | 31 | 13 |
    | 16 | 10 | 3 | 30 | | |

6._____

_____

_____

7. The following temperatures were recorded for
    seven days in Hartford, CT. Find the mean, median,
    and mode. Round to the nearest tenth if necessary.
    59°, 52°, 39°, 50°, 60°, 47°, 63°

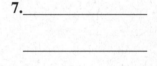

7._____

_____

_____

8. The following inches of snowfall were recorded for
    Alpine Ski Resort in one week. Find the mean,
    median, and mode.
    4 in.   5 in.   0 in.   8 in.
    14 in.   11 in.   7 in.

8._____

_____

_____

350

Chapter 1 FOUNDATIONS OF ALGEBRA

1.1 Number Sets and the Structure of Algebra

1. $\{d, a, n, c, e\}$ 3. rational   5. rational   7.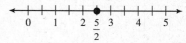

9. 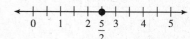   11. $\dfrac{1}{4}$ 13. $=$

15. $-12.9, \ -6.1, \ 0, \ |-1.6|, \ \left|-2\dfrac{1}{4}\right|, \ 7.4$

1.2 Fractions

1. $\dfrac{5}{16}$ 3. 12 5. 11 7. $-\dfrac{66}{126}$ and $-\dfrac{49}{126}$ 9. $2 \cdot 2 \cdot 2 \cdot 2 \cdot 2 \cdot 3$

11. $3 \cdot 5 \cdot 5 \cdot 7$ 13. $\dfrac{5}{7}$ 15. $-\dfrac{3}{5}$

1.3 Adding and Subtracting Real Numbers; Properties of Real Numbers

1. additive inverse   3. associative property of addition   5. $\dfrac{3}{2}$

7. -8 9. $-a$ 11. -7 13. -11 15. -3.8

1.4 Multiplying and Dividing Real Numbers; Properties of Real Numbers

1. associative property of multiplication   3. multiplicative identity   5. $-\dfrac{1}{7}$

7. -42.16 9. $\dfrac{1}{8}$ 11. undefined   13. 60 15. $\dfrac{2}{7}$

1.5 Exponents, Roots, and Order of Operations

1. 256 3. 0.021952 5. ± 17 7. not a real number   9. 538

11. 28.74 13. 99 15. 87

1.6 Translating Word Phrases to Expressions

1. $8c$ 3. $75+d$ 5. $3x-5$ 7. $2y-\dfrac{1}{4}$ 9. $x-7(x-16)$

11. $\dfrac{6}{x}-8$ 13. $-19+(a+b)$ 15. $x+4$

1.7 Evaluating and Rewriting Expressions

1. 24 3. -30 5. $-\dfrac{5}{6}$ 7. $-2b+10$ 9. -1 11. $-15d$

13. $-17p+47$ 15. $\dfrac{15}{11}r+\dfrac{19}{11}s$

Chapter 2 SOLVING LINEAR EQUATIONS AND INEQUALITIES

2.1 Equations, Formulas, and the Problem-Solving Process

1. no   3. yes   5. yes   7. 44 km   9. 3137.4 mi.

2.2 The Addition Principle of Equality

1. yes   3. yes   5. $-\dfrac{3}{10}$ 7. -6.8 9. 8

11. all real numbers   13. $18+x=34$; 16 mi.

2.3 The Multiplication Principle of Equality

1. 10   3. 54   5. -7 7. 7   9. $-\dfrac{2}{3}$ 11. 15   13. 26 ft.

2.4 Applying the Principles to Formulas

1. $v=m+9b$ 3. $y=\dfrac{-10x+3}{9}$ 5. $s=\dfrac{g-jh-45z}{9}$ 7. $n=\dfrac{S-Q}{by}$

9. $g^4=\dfrac{3C}{5\pi}$ 11. $p=\dfrac{9U-xc}{x}$ 13. $M=\dfrac{4}{7}(Q-21)$

2.5 Translating Word Sentences to Equations

1. $5n = -20;\ -4$ 3. $\dfrac{x}{10} = \dfrac{7}{11};\ \dfrac{70}{11}$ 5. $5x + 4 = 54;\ 10$

7. $3(b+6) = -24;\ -14$ 9. $9x - 15 = x + 1;\ 2$

11. $(2x - 9) - (x - 15) = 3;\ -3$

2.6 Solving Linear Inequalities

1. $\{x \mid x \le -1\};\ (-\infty, -1]$ 3. $\{n \mid -2 < n \le 3\};\ (-2, 3]$

5. $\{x \mid x < 5\};\ (-\infty, 5)$ 7. $\{x \mid x > -6\};\ (-6, \infty);$

9. $\{k \mid k \ge -8\};\ [-8, \infty);$ 11. $3x - 26 \ge 55;\ x \ge 27$

13. any score greater than or equal to 43

Chapter 3 PROBLEM SOLVING

3.1 Ratios and Proportions

1. $\dfrac{1}{6}$ 3. $\dfrac{19}{1}$; for every 19 miles, he uses 1 gallon of gas.   5. yes   7. no

9. -56 11. $x = 36;\ y = 25.5;\ z = 18$

3.2 Percents

1. $0.65;\ \dfrac{13}{20}$ 3. $0.14\overline{3};\ \dfrac{43}{300}$ 5. 5% 7. 287 9. 85

11. $\$85.00$ 13. $\$9.36;\ \$165.36$ 15. $\$17,850$

3.3 Problems with Two or More Unknowns

1. Amanda is 24; Michael is 4.   3.   width: 6 cm; length: 9 cm   5.   58, 60, 62

7.   14, 16, 18

3.4 Rates

1.   9 hr.   3.   faster car: 48 mph; slower car: 32 mph   5.   7 hr.

3.5 Investment and Mixture

1. $960 at 4%; $640 at 3%   3. $9000 in plan A; $11,000 in plan B   5.   4200 kg

Chapter 4 GRAPHING LINEAR EQUATIONS AND INEQUALITIES

4.1 The Rectangular Coordinate System

1. $A(-5,1)$, $B(1,2)$, $C(4,-1)$, $D(0,-3)$ 3.

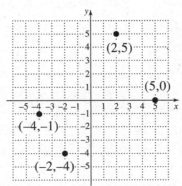

5.   IV   7.   II   9.   I   11.   nonlinear   13.   nonlinear

4.2 Graphing Linear Equations

1.   no   3.   yes

5. $(-1,-4)$, $(0,0)$, $(1,4)$

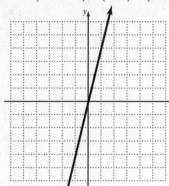

7. $(-5,-8)$, $(0,-4)$, $(5,0)$

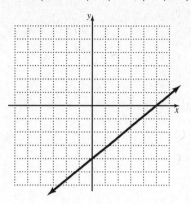

4.3 Graphing Using Intercepts

1. $(5,0),(0,6)$ 3. $(4,0)$; no y-intercept

5.

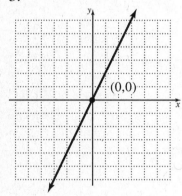

7.

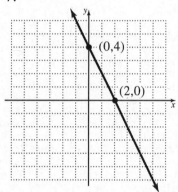

9.

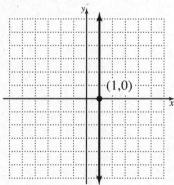

355

4.4 Slope-Intercept Form

1.

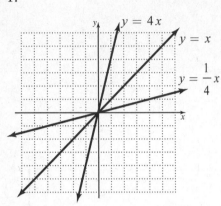

3. $m = -2; (0, 4)$

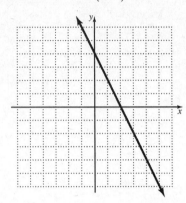

5. $y = 4x - \dfrac{1}{2}$ 7. $\dfrac{4}{3}$ 9.   1

4.5 Point-Slope Form

1. $y = -8x + 65$ 3. $y = -\dfrac{1}{2}x - \dfrac{21}{2}$ 5. $y = -\dfrac{5}{4}x + 5$

7. $7x + 4y = 47$ 9a. $y = 5x + 2$ b. $5x - y = -2$

11a. $y = -2x + 14$ b. $2x + y = 14$ 13.   parallel

4.6 Graphing Linear Inequalities

1.   yes   3.   no

5.

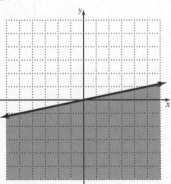

7.

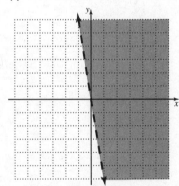

4.7 Introduction to Functions and Function Notation

1. domain: $\{7,24,38,50\}$; range: $\{2,-9,8\}$ 3. No, an element in the domain is

assigned to more than one element in the range. (5 is assigned to both 4 and 3.)

5a. 10   b. 40   c. 10   7. -2 9.

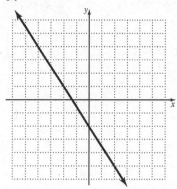

Chapter 5 POLYNOMIALS

5.1 Exponents and Scientific Notation

1. 1   3. $\dfrac{8}{125}$ 5. $\dfrac{1}{36}$ 7. $-\dfrac{1}{27}$ 9. 7,320,000

11. 0.000000143   13. 7.518×10^9

5.2 Introduction to Polynomials

1. monomial   3. coefficient: 6; degree: 7   5. trinomial; 2

7. no special polynomial name; 7   9. 5   11. 77

13. $x^5 + 8x^3 + 6x^2 + x + 7$ 15. $-22a^2p - 2p^2$

5.3 Adding and Subtracting Polynomials

1. $-x + 14$ 3. $x^2 - 8x - 2$ 5. $6x^2 + 11$ 7. $18x + 2$

9. $-10v^2 - 2v + 9$ 11. $2y^5 + 3y^4 + 5y^2 - 10$

5.4 Exponent Rules and Multiplying Monomials

1. v^9 3. $35r^3$ 5. $36q^8r^5$ 7. $7x^2$ 9. 2.63088×10^4

357

11. x^{28} 13. $-r^{22}s^{11}$

5.5 Multiplying Polynomials; Special Products

1. $-5x^2 + 15x$ 3. $-36y^6 - 16y^5 + 24y^4$ 5. $12x^2 + 5x - 3$

7. $x^3 - 6x^2 - 49$ 9. $r^4 + 8r^3 - 3r^2 - 16r + 9$ 11. $\frac{1}{2}b + 2c$

13. $9r^2 - 25$ 15. $25x^2 + 40xy + 16y^2$

5.6 Exponent Rules and Dividing Polynomials

1. 3^3 3. -7×10^{-3} 5. $-5a^3b^4$ 7. $x + 4y$ 9. $x + 1$

11. $x^2 - 9x + 81$ 13. $\dfrac{125}{64}$

Chapter 6 FACTORING

6.1 Greatest Common Factor and Factoring by Grouping

1. $1, 2, 3, 6, 9, 18$ 3. 20 5. $x(x-5)$ 7. $8m^3n^2(n+6)$

9. $-2z(3z-7)$ 11. $(7m-4)(6m-5)$ 13. $(s+3)(s+7)$

15. $(r-5t)(r+2w)$

6.2 Factoring Trinomials of the Form $x^2 + bx + c$

1. 11 3. 11 5. $(t+2)(t+4)$ 7. $(r-7)(r-10)$ 9. prime

11. $(b+3x)(b-9x)$ 13. $2x(x+2)(x+5)$ 15. $6(c-3)(c-7)$

6.3 Factoring Trinomials of the Form $ax^2 + bx + c$, where $a \neq 1$

1. $(t-5)(t-7)$ 3. prime   5. $(w-5z)(w-4z)$ 7. $2(2s-1)(s-5)$

9. $(9v-2)(2v-9)$ 11. $(7b-1)(4b+3)$ 13. $(3a+2)(3a+4)$

15. $5u(6u-7)(u+6)$

6.4 Factoring Special Products

1. $(s+6)^2$ 3. $(5a-9)^2$ 5. $(5r+7t)^2$ 7. $(2c+11)(2c-11)$

9. $(r-3)(r^2+3r+9)$ 11. $(b+4)(b^2-4b+16)$

13. $\left(5m+\dfrac{6}{5}\right)\left(5m-\dfrac{6}{5}\right)$ 15. $(y+z-2)(y^2+2yz+z^2+2y+2z+4)$

6.5 Strategies for Factoring

1. $(c-5)(a+z)$ 3. $(v+3)(v^2-3v+9)$ 5. $(r+2)(r+8)$

7. $(t+5)(t-6)$ 9. $(5x-6a)(25x^2+30ax+36a^2)$

11. $(s^2+9)(s+3)(s-3)$ 13. $(2v-9)^2$ 15. $r(3+10r)(3-10r)$

6.6 Solving Quadratic Equations by Factoring

1. $25,\ -64$ 3. $0,\ -4$ 5. $1,24$ 7. $-3,\ -4$ 9. $6,\ -\dfrac{4}{5}$

11. $5,7$ 13. 2.5 sec. and 3.5 sec.

6.7 Graphs of Quadratic Equations and Functions

1.

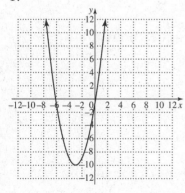

3.

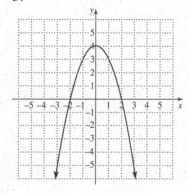

5.

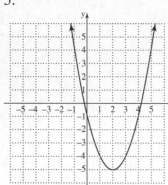

7.

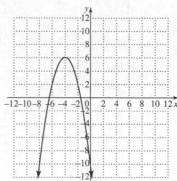

Chapter 7 RATIONAL EXPRESSIONS AND EQUATIONS

7.1 Simplifying Rational Expressions

1a. 1   b. $\dfrac{34}{9}$ 3a. $\dfrac{8}{25}$ b. undefined   5. $4, -4$ 7. $\dfrac{125u^4 z^8}{2}$

9. $\dfrac{8}{7}$ 11. $\dfrac{r+7}{r-7}$ 13. $\dfrac{a-d}{a+d}$

7.2 Multiplying and Dividing Rational Expressions

1. m^4 3. $-\dfrac{4}{7}$ 5. $\dfrac{x+2}{2x(x-2)}$ 7. $\dfrac{1}{9}$ 9. $\dfrac{w(w-4)(w+4)}{x}$

11. $\dfrac{v+4}{(v+3)(5v-3)}$ 13. 180 in.

7.3 Adding and Subtracting Rational Expressions with the Same Denominator

1. $\dfrac{x}{13}$ 3. y 5. 2   7. $z+3$ 9. $\dfrac{y-2}{y}$ 11. 2

13. $x+15$

7.4 Adding and Subtracting Rational Expressions with Different Denominators

1. $33a^6$; $\dfrac{22a}{33a^6}, \dfrac{9}{33a^6}$ 3. $12(x-2)$; $\dfrac{32x}{12(x-2)}, \dfrac{15x}{12(x-2)}$

5. $\dfrac{49}{5z}$ 7. $\dfrac{y+c}{y-c}$ 9. $\dfrac{9y-56}{(y-7)(y+5)}$

7.5 Complex Rational Expressions

1. -7 3. $\dfrac{19}{14}$ 5. $\dfrac{29}{14}$ 7. $\dfrac{y}{y-7}$ 9. $-\dfrac{1}{y(y+b)}$

7.6 Solving Equations Containing Rational Expressions

1. yes 3. 3 5. $\dfrac{63}{11}$ 7. 18 9. $\dfrac{39}{7}$ 11. $\dfrac{8}{5}$

7.7 Applications with Rational Expressions, Including Variation

1. 2 hr. 3. freighter: $\dfrac{75}{4}$ km per hr.; steamboat: $\dfrac{123}{4}$ km per hr.

5. 24 7. $\dfrac{2}{9}$ A 9. $\dfrac{33}{2}$ 11. 25

Chapter 8 MORE ON INEQUALITIES, ABSOLUTE VALUE, AND FUNCTIONS

8.1 Compound Inequalities

1. $\{x \mid -2 < x < 2\}$; $(-2,2)$

3. $\{x \mid x \geq 6\}$; $[6,\infty)$

5. $\{x \mid -4 < x < 1\}$; $(-4,1)$

7. $\{x \mid x < -6 \text{ or } x > 4\}$; $(-\infty,-6) \cup (4,\infty)$

9. $\{x \mid -4 \leq x < 6\}$; $[-4,6)$

11. $\{x \mid -4 \leq x < 3\}$; $[-4,3)$

8.2 Equations Involving Absolute Value

1. -9, 9 3. -8, 20 5. 1 7. -6, 6 9. $-\dfrac{9}{7}, \dfrac{3}{7}$

11. $-19, 11$ 13. all real numbers

8.3 Inequalities Involving Absolute Value

1. $\{x|-5 < x < 5\}$; $(-5,5)$ 3. $\{x|-5 \le x \le 1\}$; $[-5,1]$

5. $\{x|x < -2 \text{ or } x > 2\}$; $(-\infty, -2) \cup (2, \infty)$ 7. $\{\ \}$ or $\varnothing$; no interval notation

9. $\{x|x \text{ is a real number}\}$; $\mathbb{R}$

8.4 Functions and Graphing

1. domain: $\{12,\ 27,\ 34,\ 48\}$; range: $\{2,\ -10,\ 6\}$; not a function

3. domain: $\{x|x \le 1\}$; range: all real numbers; not a function   5. $\dfrac{17}{46}$ 7. 5

9.

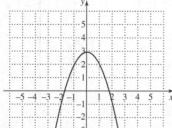

8.5 Function Operations

1. $4x^2 + 8x - 7$; $4x^2 - 8x + 7$ 3. $2x$; 4

5. $x^3 + 2x^2 + 6x + 3$; $-x^3 + 2x^2 - 6x + 3$ 7. $4x^2 + 9x - 5$; $3x + 3$

9. $6x^2 - 25x - 9$ 11. $x^4 - x^3 - 13x^2 + 31x - 18$ 13. $x-1$, $x \ne -1$

362

Chapter 9 SYSTEMS OF LINEAR EQUATIONS AND INEQUALITIES

9.1 Solving Systems of Linear Equations Graphically

1. yes   3. $(4,6)$ 5. $(1,0)$ 7. all ordered pairs along $4x - 5y = 2$

9a. consistent with dependent equations   b. infinite number of solutions

9.2 Solving Systems of Linear Equations by Substitution; Applications

1. $(3,1)$ 3. $(8,-6)$ 5. all ordered pairs along $x - 2 = y$ 7. $(-4,6)$

9. 22, 13   11. ones: 32; tens: 13   13. 431.25 km

9.3 Solving Systems of Linear Equations by Elimination; Applications

1. $(4,-1)$ 3. $(-8,6)$ 5. $(-5,7)$ 7. $(4,3)$ 9. $30°, 150°$

11. 450 mph   13. \$850 at 4%; \$1700 at 5%   15. 392 kg of Kenyan; 196 kg of Turkish   17. 4200 kg

9.4 Solving Systems of Linear Equations in Three Variables; Applications

1. yes   3. $(-2,-2,2)$ 5. $(-3,0,-4)$ 7. infinite number of solutions (dependent equations)   9. $(4,4,2)$ 11. angle A: 21°; angle B: 89°; angle C: 70°

9.5 Solving Systems of Linear Equations Using Matrices

1. $\begin{bmatrix} -1 & 6 & | & 5 \\ 3 & -5 & | & 7 \end{bmatrix}$ 3. $(-31,7)$ 5. $\begin{bmatrix} 4 & 3 & | & 1 \\ 0 & 3 & | & 9 \end{bmatrix}$ 7. $(4,-2)$ 9. $(-2,20,12)$

11. $(5,4)$ 13. $(-2,-2,0)$

9.6 Solving Systems of Linear Equations Using Cramer's Rule

1. 50   3. 140   5. $-\dfrac{4669}{13,500}$ 7. $\left(-\dfrac{6}{19}, \dfrac{55}{38}\right)$ 9. $\left(\dfrac{45}{14}, -\dfrac{155}{42}\right)$

11. $(9,-12,-12)$

9.7 Solving Systems of Linear Inequalities

1.

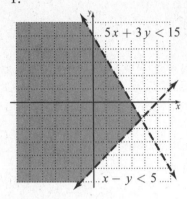

3.

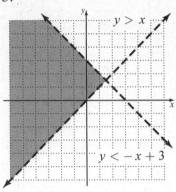

5.

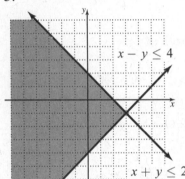

7.

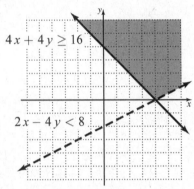

Chapter 10 RATIONAL EXPONENTS, RADICALS, AND COMPLEX NUMBERS

10.1   Radical Expressions and Functions

1. 9 3. -7 5. 2.962 7. $6x^8 y^{17}$ 9. $|x-6|$

11. $(y-2)^3$ 13. $\left\{ x \middle| x \geq -\dfrac{9}{5} \right\}$ or $\left[-\dfrac{9}{5}, \infty \right)$

10.2   Rational Exponents

1. $-\sqrt{25} = -5$ 3. $\left(\sqrt[3]{216} \right)^4 = 1296$ 5. $\sqrt[6]{(3y+x)^5}$ 7. $n^{5/6}$

9. $a^{5/2}$ 11. $-42 y^{1/6}$ 13. $\sqrt[3]{a^2}$ 15. $\sqrt[6]{y^5}$

10.3   Multiplying, Dividing, and Simplifying Radicals

1. $10pn^4$ 3. $\sqrt{\dfrac{7x}{5y}}$ 5. 7 7. $2\sqrt{10}$ 9. $3x^2\sqrt[3]{9x^2}$

11. $a^4b^3\sqrt{ab}$ 13. $5\sqrt{7}$ 15. $18xy^4\sqrt{2xy}$

10.4   Adding, Subtracting, and Multiplying Radical Expressions

1. $5\sqrt{5}$ 3. $\sqrt{5}$ 5. $-7x^2\sqrt{x}$ 7. $7+\sqrt{70}$

9. $56\sqrt{21}+21\sqrt{14}-32\sqrt{15}-12\sqrt{10}$ 11. $46+12\sqrt{10}$

13. -190 15. $72-48\sqrt{3}$

10.5   Rationalizing Numerators and Denominators of Radical Expressions

1. $\dfrac{5\sqrt{7}}{14}$ 3. $\dfrac{x\sqrt{15}}{15}$ 5. $\dfrac{5\sqrt[3]{4}}{2}$ 7. $\dfrac{2\sqrt[3]{9x}}{3x}$ 9. $\dfrac{-7-6\sqrt{7}}{29}$

11. $\dfrac{x-\sqrt{xy}}{x-y}$ 13. $\dfrac{1}{\sqrt{11}-2}$

10.6   Radical Equations and Problem Solving

1. 49 3. 112 5. 63 7. 11 9. 9 11. 4

13. $1,2$ 15. $0,256$

10.7   Complex Numbers

1. $11i$ 3. $3i\sqrt{5}$ 5. $3+9i$ 7. $-40-35i$ 9. $-12-16i$

11. $\dfrac{20}{13}-\dfrac{4}{13}i$ 13. 1 15. i

Chapter 11 QUADRATIC EQUATIONS AND FUNCTIONS

11.1   The Square Root Principle and Completing the Square

1. ±1.4 3. $\pm8i$ 5. ±6 7. $\dfrac{8}{3},-4$ 9. $-25.5,-32.5$

11. $-3, 1$ 13. $-\dfrac{5}{3}, \dfrac{1}{4}$ 15. $\dfrac{-10 \pm \sqrt{109}}{3}$

11.2   Solving Quadratic Equations Using the Quadratic Formula

1. $-3, 6$ 3. $\dfrac{1 \pm 3i\sqrt{3}}{2}$ 5.   one rational solution

7.   two nonreal complex solutions   9. $(3,0),\ (-2,0),\ (0,-6)$

11. $x^2 + 4(x+1) = 81;\ 7, 8$ 13. 5.6 sec.

11.3   Solving Equations That Are Quadratic in Form

1. $-50,\ 55$ 3. $-\dfrac{3}{2}, \dfrac{3}{2}$ 5. $1, 25$ 7. $2, 3$ 9. $\pm 4,\ \pm 5$

11. $3, 4$ 13. $\dfrac{16}{81}$ (1 is extraneous)

11.4   Graphing Quadratic Functions

1. $(7,3);\ x = 7$ 3. $(3,-3);\ x = 3$

5a. upward   b. $(-2,1)$ c. $x = -2$ d.

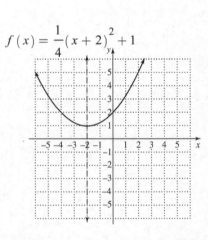

$$f(x) = \frac{1}{4}(x+2)^2 + 1$$

366

7a. $(-6,0)$, $(0,36)$ b. $h(x)=(x+6)^2$ c. upward   d. $(-6,0)$

e. $x=-6$ f.

$$h(x) = x^2 + 12x + 36$$

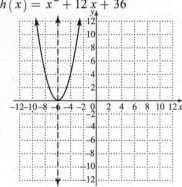

g. domain: $\{x \mid x \text{ is a real number}\}$ or $(-\infty, \infty)$; range: $\{y \mid y \geq 0\}$ or $[0, \infty)$

9a.   b. 117.551 m   c. 9.796 sec.

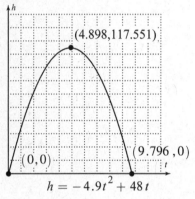

$$h = -4.9t^2 + 48t$$

11. 4:00 p.m.

11.5   Solving Nonlinear Inequalities

1. $(-3,4)$

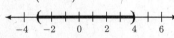

3. $(4,11)$

5. No solution or $\varnothing$

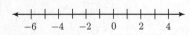

7. $(-\infty,-6)\cup(-5,2)$

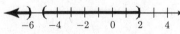

9. $(2,6]$

11. $(8,8.5)$

Chapter 12 EXPONENTIAL AND LOGARITHMIC FUNCTIONS

12.1 Composite and Inverse Functions

1. $(f \circ g)(x) = 50x^2 - 60x + 22$; $(g \circ f)(x) = 10x^2 + 17$

3. $(f \circ g)(x) = 8 - x$; $(g \circ f)(x) = \sqrt{4 - x^2}$ 5. Yes   7. No

9. $f^{-1}(x) = \dfrac{9x + 8}{1 - x}$ 11.

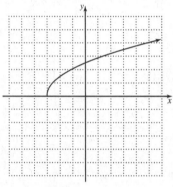

12.2 Exponential Functions

1.

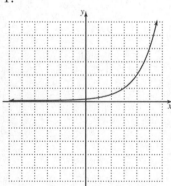

3.

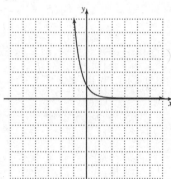

5. -4 7. $28,827.22

12.3 Logarithmic Functions

1. $\log_3 243 = 5$ 3. $\log_{1/5} 125 = -3$ 5. $10^{-5} = \dfrac{1}{100,000}$ 7. $\dfrac{1}{2}$

9. -3 11. 64   13.

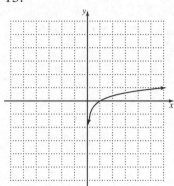

12.4 Properties of Logarithms

1. f 3. $\log_d a + \log_d q + \log_d r$ 5. $\log_3 HZ$ 7. $\log_6 13 - \log_6 7$

9. $\log_c 7$ 11. $15 \log_N t$ 13. $\log_6 7^2$ 15. $\log_b x + 7 \log_b y - 6 \log_b z$

12.5 Common and Natural Logarithms

1. 1.9685   3. 4.7412   5. 8.5129   7. 0   9. -1 11. 3.4

12.6 Exponential and Logarithmic Equations with Applications

1. 2.5850   3. 561.2160   5. 11   7. 2   9. 7

11a. $98,497.03   b. $99,334.73   13. 1.1833

Answers to Worksheets for Classroom or Lab Practice

Chapter 13 CONIC SECTIONS

13.1 Parabolas and Circles

1. opens downward; vertex: $(-2,1)$; axis of symmetry: $x=-2$ 3. $25; \left(0, -\dfrac{19}{2}\right)$

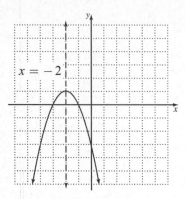

5. center: $(-2,-1)$; radius: 3   7. $(x-9)^2 + (y-4)^2 = 4$

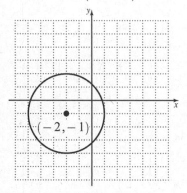

13.2 Ellipses and Hyperbolas

1.

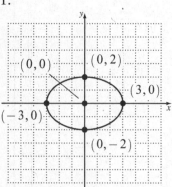

3.

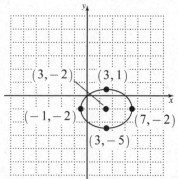

(scl = 2)

5.

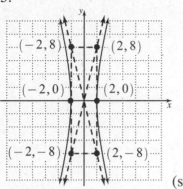

(scl = 2)

13.3 Nonlinear Systems of Equations

1. $(4,11)$, $(-7,-22)$ 3. $(6,8)$, $(-8,-6)$

5. $(6,8)$, $(6,-8)$, $(-6,8)$, $(-6,-8)$ 7. $(5,0)$ 9. $(2,5)$, $(-1,-4)$

13.4 Nonlinear Inequalities and Systems of Inequalities

1.

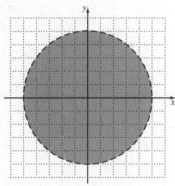

3.

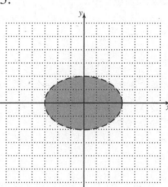

5.

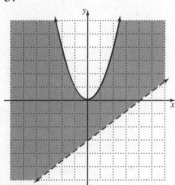

7.

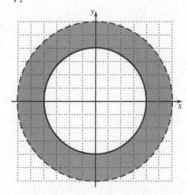

APPENDICES

Appendix A   Arithmetic Sequences and Series

1. $0, 3, 8, 15; 99$ 3. 11 5. $a_n = 2n + 5, \ a_6 = 17$

7. $-18, \ -12, \ -6, \ 0$ 9. $38, 33, 28, 23$

11. $(-2) + (-1) + 0 + 1 + 2 + 3 + 4 + 5 = 12$ 13. 1320

Appendix B   Geometric Sequences and Series

1a. 3 b. $a_{10} = 78,732$ c. $a_n = 4(3)^{n-1}$ 3a. $-1, \ 4, \ -16, \ 64, \ -256$

b. $262,144$ 5a. 3 b. $8, 24, 72, 216$ 7. -2044

9. $170\dfrac{2}{3}$ or $170.\overline{6}$ 11. $\dfrac{8}{9}$ 13. $\$819.20$

Appendix C   The Binomial Theorem

1. 2880 3. 15 5. 6 7. $x^4 + 4x^3 y + 6x^2 y^2 + 4xy^3 + y^4$

9. $729c^6 - 1458c^5 d + 1215c^4 d^2 - 540c^3 d^3 + 135c^2 d^4 - 18cd^5 + d^6$ 11. $-20x^3 y^3$

Appendix D   Synthetic Division

1. $x - 4$ 3. $2r^2 + 5r - 1$ 5. $2x^2 - 8x - 2 - \dfrac{6}{x+4}$

7. $s^3 + 5s^2 + 25s + 125$ 9. $x^2 - 10x + 10 + \dfrac{1}{x+1}$

Appendix E   Mean, Median, and Mode

1.   mean: $\$18,716.\overline{6}$; median: $\$18,700$: mode: $\$18,700$ 3.   mean: 16; median: 14;

mode: 25   5.   mean: 5.7; median: 6.2; mode: 7.9   7.   mean: $52.9°$;

median: $52°$; mode: none